1 MONTH OF
FREE
READING

at

www.ForgottenBooks.com

By purchasing this book you are eligible for one month membership to ForgottenBooks.com, giving you unlimited access to our entire collection of over 1,000,000 titles via our web site and mobile apps.

To claim your free month visit:

www.forgottenbooks.com/free54659

ISBN 978-1-5284-6349-2
PIBN 10054659

This book is a reproduction of an important historical work. Forgotten Books uses
state-of-the-art technology to digitally reconstruct the work, preserving the original format
whilst repairing imperfections present in the aged copy. In rare cases, an imperfection in
the original, such as a blemish or missing page, may be replicated in our edition. We do,
however, repair the vast majority of imperfections successfully; any imperfections that
remain are intentionally left to preserve the state of such historical works.

FISHING AND HUNTING

ON

RUSSIAN WATERS

by

D-R O. GRIMM.

ST.-PETERSBURG,
R. GOLICKE PRINTER. NEVSKY 106.
1883.

121

FISHING AND HUNTING

ON

RUSSIAN WATERS

by

D-r O. Grimm.

ST.-PETERSBURG,

B. GOLICKE PRINTER. NEVSKY 106.

1883.

Fishing and Hunting on Russian Waters

by

D-r O. GRIMM.

~~~~~~~~~

I. A Survey of the water basins of European Russia and their characteristics.

In consequence of natural conditions, the Russian Empire is not rich in sea fisheries. The inhospitable, gloomy Arctic Ocean, with its perpetual coat of ice, its short summer and almost uninhabited coast, presents no conditions favourable to the development of the fishing trade.

The Baltic and the Black Sea, in consequence of being so landlocked, and having so little salt in their water, and in consequence of their origin, are something between a sea and a fresh-water lake, with a poor, mixed fauna; besides, the representatives of the Baltic fauna are remarkable for their small size, while the Black Sea, owing to its great depth, which begins at the very coast, offers no conditions favourable to the life of most fish, seals, etc It is quite different as regards the river systems and those basins, which, though the water in them is brackish, belong, by their origin and fauna, to the category of lakes as the Caspian Sea, the Sea of Azoff and the Aral lake. These last present such favourable conditions for the life of fish, that the fisheries, existing on these lakes, not only amply recompense us for the absence of rich sea fisheries, but are preeminent among the fisheries of the whole world, as will be seen on examining the amount caught in these basins.

All Russia, especially its borders, is covered with a multitude of salt and fresh water lakes, and with a great number of rivers. In European Russia alone there are:

The Caspian Sea, which covers an area of				8413,25	sq. miles
The Sea of Azoff.	»	»	»	»	637,70	»
Lake Ladoga . . .	»	»	»	»	336,60	»
Lake Onega . . .	»	»	»	»	228,39	»
Lake Peipus and the Pskoff lake . .	»	»	»	»	65,99	»
Lake Beloš . . .	»	»	»	»	21,40	»
Lake Ilmen . . .	»	»	»	»	16,79	»
Lake Vetch . . .	»	»	»	»	19,10	»
The Koobinsky lake	»	»	»	»	7,90	»
Lake Seliger . . .	»	»	»	»	3,62	»

and there are several thousand small lakes, which, taken together, occupy an area of no less than a thousand square miles.

There are 84 navigable rivers in European Russia, which, taken together, are 30,106 versts long, and several of them are 5 versts wide, not counting the mass of separates mouths, which cover a great area.

Although these basins have animals common to all of them, they also have their special representatives, forming the characteristics of their fauna, which show the origin or history of the basins themselves.

European Russia is divided into the following basins.

1) *The basin of the Arctic Ocean* and *White Sea,* having a fauna characteristic of the Arctic zone, with a multitude of *Amphipoda* and *Isopoda* as representatives of the *crustacea, Salmoniqae* and *Gadidae* of the fish, and *Pinnipediā* and *Cetācea* of the mammalia.

2) *The basin of the Baltic,* counting the Baltic itself and the rivers falling into it, and the lakes—Ladoga, Onega, Ilmen, Peipus and a multitude of small lakes. This basin is of newer formation and occupies approximately the area of glaciers of the so called glacial period. The Baltic was for some time connected with the White Sea by the flow of water from lake Onega to the north. That is the reason that some of the White Sea animals have found their way into the Baltic.

However, these form only about 5%, while 95% of the Baltic fauna have come from the German Ocean and the Atlantic. The Baltic has no independent fauna of its own—only some unimportant varieties (e. g. Tellina solidula var. baltica) have been formed, not to speak of the many cases of diminution in the size of animals, which, it is true, has caused them to be classed as independent forms (e. g. Clupea harengus var. membrus, Asmerus eperlanus var. spirinchus, Musis aculata var. relicta), but only by persons who want to found a new species. As regards the other Russian basins, the most characteristic animals of the Baltic is the broadclawed craw-fish (Astacus fluviatis) and the eel (Anguilla vulgaris)

3) *The basin of the Black Sea* is divided, by the nature of its fauna, into two distinct parts — a) *the Black Sea* proper, which has a Mediterranean fauna, the members of which have penetrated (and continue to penetrate) into the Pontus through the sea of Marmora, and pass chiefly along the coasts of Asia Minor towards the east. — b) *the basin of the Azoff and Kherson,* — which consists of the Sea of Azoff and the northwest corner of the Black Sea, into which the Dnieper, the Bug and Dniester fall. This basin is divided by the Crimean peninsula into two parts, which are more or less connected with the Black Sea, and which have borrowed from it its Mediterranean fauna; but as they contain salt water, they have retained to the present day the fauna of the Ponto-Aralo-Caspian basin, to which they once belonged.

4) *The basin of the Caspian* is the greatest as regards the area occupied by its river system, and at the same time the most important as regards its fisheries.

To understand the constitution of the Caspian fauna, one must look at the origin of this basin.

The Sarmatian basin, which existed in the miocenic epoch, occupied the great area of what is now the south of Russia and central Europe, and consisted of many connected basins; during the pliocenic period it became considerably smaller, and was reduced to the still smaller postpliocenic basin, which maybe called the Ponto-Aralo-Caspian. Perhaps this basin was divided by the Caucasus-Crimean chain of mountains, which lay in the direction of the Danube; this is seen by the difference of the fauna not only of the Sea of Azoff, out also of the Cherson part of the Black Sea.

Besides the above mentioned parts of the Black Sea, this basin embraced the Caspian Sea, which was connected with the Sea of Azoff by a narrow channel running along the valley of the river Manitch, and reaching along the Volga almost as for as Samara.

Towards the south-east this basin, rounding the peninsula of Mantchishlon from the south, turned to the north, embracing the whole of the Aral Sea, and stretched to the east of the Ural and the Arctic Ocean.

But the connection of this salt water Cherson-Aralo-Caspian basin with the Arctic Ocean was soon broken, and the basin became considerably smaller from the Asiatic side. The receding of the waters from the eastern coast, in consequence of the rise of the mainland, produced the land-locked bays of the Caspian Sea, with their great evaporation and absence of fresh-water tributaries.—In consequence of these last two causes, these bays, which were replenished with the brackish water of the Caspian, became very salt, and part turned into salt lakes (Elton, Baskoontchak etc.), while part, namely the largest of them Aral, became so salt that only those animals remained in it, that were capable of living in such water (we find the Aral fauna in the salt bays of the East Caspian). This part, however, by a change in the course of the Amou-Daria river, again received a supply of fresh water, became enlarged, and now, though its waters are less salt than those of the Caspian Sea, still retains the poor fauna of the latters salt bays. This process of the formation of a salt lake out of a bay of the Caspian is to be seen at the present day—Lake Karaboogaz, whose water is surcharged with salt, and which receives from the Caspian about 32,000.000 poods of salt daily.

With the rising of the eastern bank, the waters of the Caspian fell towards the north, in the valley of the Volga and the Ural, but for a long time this basin was still connected with the Sea of Azoff by the Kooma-Manitch system.

All these grades of development of the Caspian, the Sea of Azoff and the Sea of Aral are strongly marked in their fauna, as the investigation of the Caspian Sea showed. In the Caspian Sea to the present day we find, in great depths, forms which characterize the so-called Sarmatian layer of the miocenic period, as the molluscs *Dreyssena Brardii*, *Dreyssena rostriformis*, *Cardium catillus*, *Planorbis micromphalus* (to the depth of 150 fathoms).

At the depth of 150—160 fathoms we find representatives of the Aralo-Caspian layer of the postpliocenic period, as the *Cardium pseudo-catillus*, *Cardium Baerii*, members of the genus *Adacna*, *Dreyssena Caspia* etc. At the depth of 15 fathoms, we only find pristine forms, and those which have been developed at the latest period; these are new forms, which have adapted themselves to the new conditions of life, as the *Cardium edule*, *Cardium longipes*, *Neritina litturata*, *Hydrobia stagnalis*.

Of the other types of the animal Kingdom, which have left no trace of their existence in the layer of the above mentioned geological epochs, in the Caspian we find the following typical groups.

Of the sponges we have members of the Renieridæ family. Of the worms, with the exception of fresh and salt water animals, we only find a member of the leech family (*Archiobdella Esmonti*) and some representatives of the genus *Amphicteis*.

Of the lichens we have Laguncula and Bowerbankia.

Of the crustacea, besides some lake Copepoda (as, for instance, the highly interesting *Bythotrephes socialis*), there are as many as a hundred species of the *Gammaridae*, some *Mysidae, Idotea entomon* and *Cumacea* not to mention the *Astacus leptodactylus* and *Astacus pachypus*, of which the last distinctly characterizes the Cherson-Caspian basin.

Of the fish, the most characteristic are the *Gobiidae*, as indigenous, *Cyprinidae*, as emigrants from Central Asia, the *Acipenseridae* and *Salmonidae*, as emigrants from the north. The seal *(Phoca Caspia)* ought, of course, to be classed with the northern emigrants.

Thus the Caspian Sea, in its present fauna, has the characteristics of a brackish lake of the tertiary period, enriched by emigrants chiefly from the north.

As regards the multitude of various amphipodae, belonging to the family of Gammaridae, the Caspian fauna is analogous to that of the Baikal and of Lake Superior in North America.

Although the river systems, falling into these basins, are quite independent of each other, and have a typical fauna of their own, they are connected with each other near their sources by small rivers and lakes, through which, as also through artificial communication (canals), those animals that go up as high as the source of the river, can pass from one system into another.

In this manner, the sterled has of late emigrated from the Volga through the river Sheksna, Prince Würtemburg's Canal, the Koobinsky lake into the Sookhona and the North Dvina. The lake variety of the smelt (or sparling) has found its way from the basin of the Baltic, out of lake Vela, through the connecting canal into Lake Shlino and through the river Shlina into Lake Metino. It sometimes happens that the eel finds its way from the Baltic into Lake Seliger and the river Don.

II. A list of the fish of European Russia, with their geographical destribution.

1. Gasterosteus aculeatus L.—In the Arctic Ocean, the White Sea and the Baltic, with the rivers falling into them; all the way to Kamtchatka. They are rarely to be found in the rivers of the Black Sea.

2. Gasterosteus pungitius L. — In the same places.
3. Gasterosteus platygaster Kessl. — In the Black Sea, the Caspian and the Aral, and the mouths of their rivers.
4. Gasterosteus spinachia L. — In the Baltic, to Cronstadt.
5. Perca fluviatilis L. — Everywhere to the river Lena. It is, however, not found in the river Koora, though in the Caspian it is everywhere.
6. Acerina cernua L. -- Everywhere to the river Lena, with the exception of the south part of the Caspian, and the rivers falling into it (r. Koora), and of the Aral Sea.
7. Acerina rossica Cuv. — It is found only in the lower parts of the rivers which fall into the Black Sea and the Sea of Azoff.
8. Percarina Demidoffi Nordm. — In the Black Sea, at the mouths of the Dniester, Bug and Dnieper.
9. Sebastes norvegicus. — Arctic Ocean.
10. Lucioperca volgensis Pall. — In the rivers (chiefly in the lower parts) which fall, from the north, into the Black, Azoff and Caspian seas. Mostly in the Volga; in the Caspian as far as Baku.
11. Lucioperca sandra Cuv. — In all large rivers and lakes of the Black Sea, Azoff, Caspian, Aral and Baltic basins, as far as $66^1/_2°$ north. lat.
12. Lucioperca marina Cuv. — In the Caspian, the Sea of Azoff, and in the less salt parts of the Black Sea (not going up the river).
13. Aspro zingel L. —- In the Dniester.
14. Serranus scriba L. — Rarely to be found in the Black Sea; along the coasts of the Crimea and Abhasia.
15. Serranus cabrilla L. — Rarely to be found in the Black Sea along the Anatolian coast.
16. Dentex rivulatus Bennet. — Ditto.
17. Smaris chryselis Cuv. — In the Black Sea near the Crimea.
18. Mullus barbatus L. — In great quantities in the Black Sea, along the coasts of the Crimea and Abhasia.
19. Mullus surmuletus L. — Is occasionally found in the Black Sea and the Baltic.
20. Sargus annularis L. — Found in great quantities in the Black Sea.
21. Pazellus erythrinus. — In the Black Sea, along the south and south-east coasts.

22. Charanx puntazzo L. — In the Black Sea (rare).
23. Scorpaena porcus L. — Often found in the Black Sea.
24. Sebastes norvegicus Fr. M. — Arctic Ocean, along the Mourman coast.
25. Cottus gobio L. — Found everywhere with the exception of Trans-caucasia and the extreme north, namely in the Petchora river as far as 67° north lat.
26. Cottus poecilopus Heckel. — In the government of Olonetz.
27. Cottus scorpius L. — In the Arctic Ocean, White Sea and the Baltic, but not in the rivers.
28. Cottus quadricornis L. — All along the coast of Siberia in tho Arctic Ocean, as far as the White Sea, in the White Sea itself, in the Baltic and Lake Ladoga.
29. Scelus hamatus Kröy. — In the Arctic Ocean near the Mourman coast, and in the White Sea.
30. Agonus decaganus Bl. — In the Arctic Ocean near the Mourman coast.
31. Trigla hirundo Bl. — In the Black Sea, near the Crimea.
32. Trigla poeciloptera Cuv. — In the Black Sea, along the southern coast of the Crimea (rare).
33. Trigla cuculus Bl. —   Ditto.
34. Uranoscopus scaber L. — Often caught in the Black Sea on the Crimean coast; rarely found at Odessa.
35. Trachinus draco E. —   Ditto.
36. Umbrina cirrhosa L. — In the Black Sea, near the Crimea o and the Caucasus.
37. Corvina nigra Bl. —   Ditto.
38. Scomber scomber L. — Found in great quantities in the Black Sea, but does not enter the Sea of Azoff. In the Baltic as far as the Gulf of Finland.
39. Thynnus thynnus L. — Found singly, along the southern coast of the Black Sea.
40. Pelamys saida Bl. — In the Black Sea, in small shoals; goes as far as the Crimea.
41. Zeus pungio Cuv. — One was caught in the Black Sea near Sebastopol.
42. Trachurus trachurus L. — In considerable quantities in the Black Sea Rarely found in the Baltic.
43. Trachurus lacerta Ball. — In the Black Sea.
44. Temnodon saltator Cuv. — In the Black Sea.
45. Xiphias gladius L. — In the Black Sea and the Baltic, but rarely.
46. Gobius minutus Penn. — In the Baltic Sea, near Libau.

47. Gobius niger. — In the Baltic Sea, near Helsingfors.

48. Gobius Eckströmi Günth. —In the Baltic Sea, near Libau.

49. Gobius jozo, G. — In the Black Sea, along the coasts of the Crimea.

50. Gobius quadricapillus Pall.   Ditto.

51. Gobius amphiocephalus Pall. — In the Black Sea, from Odessa to Kertch.

52. Gobius capito Cuv. — In the Black Sea, from Lebastopol to Kertch.

53. Gobius capitonellus Kessl. — In the Black Sea.

54. Gobius albosignatus Kessl. — In the Black Sea.

55. Gobius marmoratus Pall. — In the Black Sea and the Caspian, goes up the rivers.

56. Gobius blennioides Kessl. — In the Caspian Sea, in the Bay of Baku.

57. Gobius macropterus Nordm. — In the Black Sea.

58. Gobius semipellucidus Kessl. — In the Caspian, Bay of Astrabad.

59. Gobius lugens Nordm. — River Kodar, falling into the Black Sea.

60. Gobius macropus. De Filippi. — In Lake Paleostom, near the Black Sea.

61. Gobius exanthematosus Pall. — In the Black Sea, along the coast of the Crimea.

62. Gobius constructor Nordm.—Near Kertch, and in the small rivers of Abasia, Mingrelia and Georgia.

63. Gobius ratan Nordm.—In the Black Sea, from Odessa to Kertch.

64. Gobius paganellus L.—In the Black Sea.

65. Gobius Goebelii Kessl.—In the Caspian, at Baku.

66. Gobius Bucchichi Steind. — In the Black Sea, from Sebastopol, to Theodosia.

67. Gobius melanostomus Pall.—Found in great quantities in the Black Sea, the Sea of Azoff and the Caspian; it goes up pretty high into the rivers,—up the Dniester as far as Mogileff, up the Dnieper as far as Ekaterinoslav, up the Volga as far as Sarepta.

68. Gobius irriscens Pall.—In the Black Sea.

69. Gobius caspius Eichn.—In the Caspian, chiefly in the south.

70. Gobius bathybius Kessl.—In the Caspian, at the depth of 108 fathoms.

71. Gobius fluviatilis Pall.—In the Black Sea, Sea of Azoff, the Caspian and in the rivers falling into these seas.

72. Gobius Bogdanowii Kessl.—In the Caspian.

73. Gobius cephalarges Pall.—In the Black Sea.

74. Gobius platyrostris Pall.—In the Black Sea.

75. Gobius cyrtus Kessl.—In the river Koora.
76. Gobius Weidemanni Kessl.—From Trans-Caucasia (?).
77. Gobius Kessleri Günth.—In the Black Sea and the Caspian, enter the mouths of the rivers.
78. Gobius eurycephalus Kessl.—In the Black Sea.
79. Gobius eurystomus Kessl.—In the southern parts of the Caspian, at Baku and Krasnovodsk.
80. Gobius syrman Nordm.—In the Black Sea, from Odessa to Kertch.
81. Gobius Trautvetteri Kessl.—In the Black Sea.
82. Gobius batrachocephalus Pall.— In the Black Sea.
83. Gobius gymnotrachelus Kessl.—In the rivers Dniester, Bug and Dnieper.
84. Gobius Burmeisteri Kessl.— The mouth of the river Rion.
85. Gobius macrophthalmus Kessl. — In the Caspian Sea, at the depth of 9—20 fathoms.
86. Gobius nigronotatus Kessl.—In the Caspian Sea, at the depth of 20 fathoms.
87. Gobius cobitiformis Kessl.—The Bay of Sebastopol in the Black Sea.
88. Gobius leopardinus Nordm.—In the Black Sea.
89. Gobius lencoranicus Kessl.—In Lencoran, in a bog.
90. Gobius longecaudatus Kessl. — Three varieties of this species have been found in the Caspian Sea at the depth of from 44 — 220 fathoms.
91. Gobiosoma caspium Kessl. — Found in the Caspian at the depth of 9—20 fathoms.
92. Latrunculus pellucidus Nardo.—In the Black Sea, near Odessa.
93. Benthophilus macrocephalus Pall.—In the Black Sea, chiefly at the mouth of the river Dnieper, Bug and Dniester, it is found in all parts of the Caspian, but chiefly in the south; at the depth of from 7—20 fathoms.
94. Benthophilus leptocephalus Kessl.—In the Caspian Sea, at the depth of from 39—108 fathoms.
95. Benthophilus ctenolepidus Kessl.—In the Caspian Sea, at the depth of from 7—35 fathoms.
96. Benthophilus spinosus Kessl. — In the Caspian, at the depth of 20 fathoms.
97. Benthophilus Bærii Kessl. — In the Caspian Sea, at the depth of 37—38 fathoms.
98. Benthophilus leptorhynchus Kessl.—In the Caspian Sea, at the depth of 70 fathoms.

99. Benthophilus granulosus Kessl. — In the Caspian, at the depth from 20 feet to 20 fathoms.
100. Benthophilus Grimmi Kessl. — In the Caspian Sea at the depth of 32—108 fathoms.
101. Callionymus festivus Pall.—In the Black Sea, chiefly along the coast of Crimea.
102. Cyclopterus lumpus L. — In the Arctic Ocean and White Sea. Found in every fact of the Baltic, but not often.
103. Liparis vulgaris Flem.—In the Arctic Ocean on the Mourman Coast, and in the Baltic along the Coasts of Finland.
104. Liparis montagui Donov. — In the White Sea.
105. Lophius piscatorius L.—Comes into the Black Sea and the Baltic.
106. Anarrhichas lupus L.—In the Arctic Ocean and the White Sea.
107. Blennius gattorugine Brünn. — In the Black Sea, at Sebastopol and Theodosia.
108. Blennius tentacularis Brünn.—In all parts of the Black Sea.
109. Blennius sanguinolentus Pall.—In the same place, in great quantities.
110. Blennius sphinx Cuv.   Ditto.
111. Blennius pavo Kisso.   , Ditto.
112. Blennius galerita L.   Ditto.
113. Blennius melanio Kessl.—In the Bay of Sebastopol, in the Black Sea.
114. Blennius macropteryx.—In the Bay of Salta, in the Black Sea.
115. Stichæus medius Reinh.—The White Sea.
116. Stichæus punctatus Fabr.—The White Sea.
117. Centronotus gunellus L. — The Arctic Ocean, Mourman Coast, also the White Sea and the Baltic.
118. Zoarces viviparus Schoner.   Ditto.
119. Atherina pontica Eichn.—Found in great quantities in the Black Sea and the Caspian,
120. Atherina hepsetus L.—In the Black Sea.
121. Mugil cephalus Cuv.—In the Black Sea and the Sea of Azoff; in the Baltic as far as the coast of Livonia.
122. Mugil chelo Cuv.—In the Black Sea.
123. Mugil auratus Risso.   Ditto.
124. Mugil saliens Risso.—In the Black Sea, and the Sea of Azoff; finds its way into the salt Shabolot lake.
125. Lepadogaster Gonanii Lacip.—In the Black Sea.
126. Lepadogaster Decandolii Risso.   Ditto.
127. Lepadogaster bimaculatus Penn.   Ditto.

128. Heliastes chromis Cuv.   Ditto.

129. Labrus turdus L.   Ditto (rarely).

130. Labrus prasoctilhes Pall.   Ditto.

131. Crenilabrus pavo Brünn.   Ditto (near the Crimea).

132. Crenilabrus grisens L.   Ditto.

133. Crenilabrus quinquemaculatus Bl.   Ditto.

135. Crenilabrus ocellatus Forsk. — Found in great quantities in the Black Sea along the coasts of the Crimea and the Caucasus.

136. Crenilabrus rostratus Bl. — In the Black Sea at Sebastopol.

137. Ctenolabrus cinereus Pall. — In the Black Sea (rarely).

138. Coris julis L. — In the southern parts of the Black Sea.

139. Ammodytes Tobianus L. — In the Arctic Ocean, Mourman Coast.

140. Gadus morrhua L. — In the Arctic Ocean, White Sea, and the Baltic (in which it goes as far as Cronstadt).

141. Gadus nawaga Köbr. — In the White Sea and to the west of it in the Arctic Ocean.

142. Gadus saida Lepech. — In the Arctic Ocean.

143. Gadus aeglefinus L. — In the Arctic Ocean, Mourman Coast.

144. Gadus viress L. —   Ditto.

145. Gadus euxinus Nordm. — Found in great numbers in the Black Sea.

146. Lota vulgaris Cuv. — Found everywhere in fresh water, with the exception of the tributaries of the south Caspian.

147. Motella tricirrata Bl. — In all parts of the Black Sea.

148. Molva vulgaris Flemm, — In the Arctic Ocean.

149. Ophidium barbatum E. — In the Black Sea.

150. Brosmius brosme Müll. — In the Arctic Ocean.

151. Macrurus rupestris Fbr. — In the Arctic Ocean.

152. Rhombus maximus L. — In the Arctic Ocean. Goes as far as Cronstadt in the Baltic.

153. Rhombus maeoticus Pall. — Found in considerable quantities in the Black Sea and the Sea of Azoff.

154. Rhombus laevis Bond. — It comes from the Mediterranean into the Black Sea, and from the German Ocean into the Baltic, but seldom.

155. Pleuronectes platessa L. — In the Arctic Ocean and White Sea.

156. Pleuronectes flesus L. — In the Arctic Ocean, White Sea, Baltic, the Black Sea and the Sea of Azoff. Goes up into the rivers and lakes.

157. Pleuronectes dwinensis Liljeb. — In the White Sea.
158. Pleuronectes cynoglossus L. — In the Arctic Ocean.
159. Pleuronectes limanda L.   Ditto.
160. Hippoglossoides limandoides Bl.   Ditto.
161. Hippoglossus vulgaris Flem.   Ditto.
162. Solea nasuta Pall. — Found everywhere in the Black Sea in great quantities, also in the Sea of Azoff.
163. Silurus glanis L. — Found in the rivers of the Baltic, Black Sea, Sea of Azoff, Caspian and Aral, and in those parts of these seas which do not contain much salt.
164. Salmo salar L. — In the Arctic Ocean, White Sea, Baltic, and in the Ladoga and Onega lakes.
165. Salmo labrax Pall. — In the Black Sea.
166. Salmo trutta L. — In the White Sea and the Baltic.
167. Salmo caspius Kessl. — In the Caspian Sea; ascends only the Koora river.
168. Salmo salvelinus L. — In Lake Ladoga and Lake Onega.
169. Salmo alpinus. — In the Arctic Ocean, near the Mourman Coast and Nova Zembla.
170. Salmo ischchan Kessl. — In Lake Gaktcha, in the Caucasus.
171. Salmo gegarkuni Kessl.   Ditto.
172. Salmo fario L. — In the smaller tributaries of the White Sea, the Baltic, the Black Sea and the Caspian.
173. Thymallus vulgaris Nils. — In the small rivers, and the upper parts of the big rivers, of the Arctic Ocean, White Sea, Baltic, Black Sea and Caspian.
174. Luciotrutta leucichthys Güld. — Found in the north of the Caspian. From there it ascends the Volga and the Ural.
175. Coregonus nelma Pall. — In Lake Koobin, North Dvina and in the Siberian rivers.
176. Coregonus polkur Pall. — In the rivers Mezen and Petchora.
177. Coregonus omul Lepech.   Ditto.
178. Coregonus peled Lepech. — In the Petchora.
179. Coregonus albula L. — In the Arctic Ocean, White Sea and the Baltic, and in the lakes belonging to these systems.
180. Coregonus fera Jurine. — In Lake Ladoga and Lake Onega.
181. Coregonus Baerii Kessl.   Ditto.
182. Coregonus Nilssonii. — Ditto (also in Lake Koobin).

183. Coregonus Widegreni Malmgren. — In the northern parts of Lake
Ladoga.
184. Coregonus lavaretus L. — In the Baltic Sea.
185. Coregonus maraena Bloch. — In lake Peipus, and in the lakes in
the Kingdom of Poland.
186. Coregonus tcholmugensis Danilewski. — In lake Onega, in the Tcha-
moush bay.
187. Osmerus eperlanus L. c. var. spirinchus. — In the White Sea and
the Baltic, and the lakes belonging these
systems.
188. Mallotus villosus Müll. — In the Arctic Ocean.
189. Esox lucius L. — In all the rivers of Russia in Europe and in
Siberia, with the exception of Transcaucasia
and the Crimea.
190. Belone acus Cuv. — In the Baltic as far as Cronstadt and near
the northern coasts of the Black Sea.
191. Cyprinus carpio L. — Found in great quantities in the sea of Azoff
and the Caspian, and in these parts of the
Black Sea which are not very salt; also in
all the rivers falling into these seas. It is
wanting in the basins of the White Sea, the
Baltic and the Arctic Ocean.
192. Carassius vulgaris Nilss. c. var. gibelio. — Found everywhere in
fresh water, as far as about 65° n. lat.
193. Capoeta fundulus Pall. — In the river Koora and its tributary Araks.
194. Capoeta Sevangi De. Filippi. — In Lake Gaktcha.
195. Capoeta Hochenackeri Kessl. — In the river Araks.
196. Barbus vulgaris Flemm. — In the rivers of the Sea of Azoff and
the Black Sea.
197. Barbus tauricus Kessl. — In the small rivers of the Crimea.
198. Barbus bulatmai Gmel. — In the Caspian (chiefly in the south) from
there it enters the Koora. Found also in the
Aral Sea.
199. Barbus ciscaucasicus Kessl. — In the rivers Kooma, Terek and
Soonsha, which fall into the Caspian.
200. Barbus caucasicus Kessl. — In the river Koora.
201. Barbus goktschaikus Kessl. — In Lake Gaktcha.
202. Barbus cyri De-Filippi. — In the river Koora.

203. Barbus brachycephalus Kessl. — Found in great quantities in the south Caspian, where it enters the Koora. Also found in the Aral Sea.

204. Barbus mursa Güld.—In the tributaries of the Koora.

205. Barbus mursoides Kessl —In the Koora.

206. Gobio fluviatilis Rond.—Found in all fresh water.

207. Gobio uranoscopus Agass.—In the systems of the Volga, the Dniester, Kooma and Koora.

208. Leuciscus rutilus L. — In all fresh water basins, and in great quantities in the Sea of Azoff and the Caspian.

209. Leuciscus Frisii Nordm. — In these parts of the Black Sea which are less salt, in the Sea of Azoff and the Caspian; in these last two places it is found chiefly in the central and southern parts, from these it ascend the Koora and the Terek.

210. Squalius cephalus L.—In all European Russia, with the exception of Trans-Caucasia.

211. Squalius turcicus De Filippi.—In the Koora and Araks.

212. Squalius borysthenicus Kessl.—In the Dnieper.

213. Squalius leuciscus Heckel. — In all European Russia; none in the Caucasus.

214. Squalius Danilewskii Kessl. — In the rivers falling into the Sea o Azoff.

215. Idus melanotus Heckel· — All over European Russia, as far as Petchora; none in the Caucasus.

216. Scardinius erytrophthalmus L.—Found everywhere, with the exception of the extreme north.

217. Phoxinus lævis Agass.—Found in all swift mountain rivers and brooks, often in great quantities.

218. Tinca vulgaris Cuv.—Found everywhere as far as 62° n. lat.

219. Chondrostoma nasus L.—In the river systems of the Sea of Azoff, the Black Sea and the Caspian, with the exception of Trans-Caucasia; also in the rivers of the South Baltic.

220. Chondrostoma variabile Jakowlew. — At the mouths of the Volga and the Ural.

221. Chondrostoma oxyrinchum Kessl.—In the Kooma and the Terek.

222. Chondrostoma cyri Kessl. — In the upper parts of the Koora and in the small rivers of western Trans-Caucasia which fall into the Black Sea.

223. Chondrostoma regium Heckel.—In the upper parts of the Araks.

224 Rhodeus amarus Bloch.—Found everywhere as far as Eastern Siberia, excepting the basins of the White Sea and the Aral.

225. Abramis brama L. — Found everywhere in fresh water and in the Sea of Azoff, the Caspian and the Aral Sea.

226. Abramis ballerus L.—In the river systems of the Baltic, the Black Sea, the Sea of Azoff and the Caspian, also in the fresher parts of these seas.

227. Abramis sopa Pall. — In the rivers and fresher parts of the Black Sea, Sea of Azoff, the Caspian, and Aral, with the exception of the rivers of Trans-Caucasia. In the river Volchoff of the Baltic system.

228. Abramis vimba L. — In the rivers and fresher parts oj the Baltic, the Black Sea, and the Sea of Azoff.

229. Abramis persa Gm.—In the Caspian (south) and in the fresher parts of the Black Sea; from these seas it ascends the rivers of Trans-Caucasia and the Crimea.

230. Abramis microlepis De Filippi.—The upper parts of the Koora.

231. Blicca björkna Artedi.—All over Europ. Russia, in fresh water, in Finland only as far as 62° n. lat.

232. Abramidopsis Leuckartii Heckel.—Found in all rivers (singly).

233. Bliccopsis abramo-rutilus Holandre.   Ditto.

234. Aspius rapax Leske.—In the rivers of the Baltic, Black Sea, Caspian and Sea of Azoff, and the fresher parts of these seas.

235. Aspius erytrostromus Kessl.—Found in the Caspian, whence it ascends the Koora and Sefid-Rood; also in the Aral Sea and its rivers.

236. Aspius hybridus Jacowlew.—Mouth of the Volga.

237. Alburnus lucidus Heckel.—In all fresh waters.

238. Alburnus chalcoides Güldenst.—In the Black Sea, Caspian, Azoff and Aral seas; ascends the rivers.

239. Alburnus tauricus Kessl.—In the river Salgir, Crimea.

240. Alburnus iblis Heckel.—Upper part of the Araks.

241. Alburnus Filippi Kessl.—-Upper part of the Koora.

242. Alburnus punctulatus Kessl.   Ditto.

243. Alburnus Hohenackeri Kessl.—Mouth of the Koora.

244. Alburnus fasciatus Nordm.—In the mountain brooks of the Crimea, the Caucasus and Turkestan.

245. Alburnus bipunctatus Bloch.—Found in the rivers Visla, Nieman, Vilia, Dnieper, Desna, Vorskla, Slutch, Oka and Soonsha, i. e. in the river systems of the Baltic, Black Sea and the Caspian.

246. Leucaspius delineatus Heckel. — In brooks of the Baltic, Black Sea and the Caspian.

247. Pelecus cultratus L.—In the Baltic, Black Sea, Azoff, Caspian, and Aral Sea, from which it goes very far up the rivers, and even partly lives in them; it is not to be found in Finland, or north of Lake Ladoga.

248. Misgurnus fossilis L.— Found everywhere in fresh water, excepting the basins of the White Sea and Trans-Caucasia.

249. Nemachilus barbatus L. — Everywhere, excepting Trans-Caucasia, where it is replaced by the species:

250. Nemachilus Brandtii Kessl.—

251. Cobitis taenia L.—Everywhere in fresh water, (? In the basin of the White Sea) in Trans-Caucasia it is replaced by the species.

252. Cobitis Hohenackeri Brandt and

253. Cobitis aurata De Filippi.

254. Cobitis caspia Eichw.—In the Caspian Sea.

255. Engraulis encrasicholus L.—Found in great quantities in the Sea of Azoff and the Black Sea.

256. Alosa pontica Eichw. and

257. Alosa caspica Eichw.—Are found in the Black Sea, Sea of Azoff and Caspian. From these seas the yascend the Aral, Volga, Don, Dniester and Dnieper in great quantities.

258. Clupea cultiventris Nordm. — In the north of the Black Sea, goes up the estuaries of the Dnieper and Dniester in great quantities.

259. Clupea delicatula Nordm. — Found in great quantities in the Black Sea and the Caspian, but it is not caught.

260. Clupea harengus L.—In the White Sea and the Baltic.

261. Clupea pilchardus Pall.— A few find their way into the Black Sea.

262. Clupeonella Grimmi Kessl.—In the Caspian, in great numbers.

263. Maletta vulgaris Val.—In the Baltic Sea.

264. Anguilla vulgaris Turton.—In the basin of the Baltic.
265. Conger vulgaris Cuv.—In the Black Sea and the Baltic (but rarely).
266. Siphonostoma typhle L.—In the Baltic, Black Sea and Sea of Azoff.
267. Syngnathus acus L.—In the Black Sea.
268. Syngnathus tenuirostris Rathke.—In the Black Sea.
269. Syngnathus bucculentus Rathke. — In the Black Sea, Sea of Azoff, and Caspian, entering the mouths of the rivers.
270. Nerophis ophidion L.—In the Baltic and the Black Sea.
271. Nerophis aequoreus L.—In the Baltic Sea.
272. Hippocampus antiquorum Leachr.—In the Black Sea and the Baltic.
273. Acipenser ruthenus L. In the rivers of the Black Sea, Sea of Azoff, Caspian and White Sea, and also the Siberian rivers of the Arctic Ocean.
274. Acipenser Güldenstaedtii Brandt.—Black Sea, Azoff, Caspian and their rivers; and rivers of north Siberia.
275. Acipenser sturio L —Baltic and Black Sea and their rivers.
276. Acipenser schypa Lovetzky.—In the lack Sea, Azoff, Caspian and Aral, and in their rivers.
277. Acipenser huso L.—In the Black Sea, Azoff and Caspian and in their rivers.
278. Acipenser stellatus Pall.   Ditto.
279. Acanthias vulgaris Risso. — In the Arctic Ocean and Black Sea in considerable quantities, and forms an article of commerce. Rarely met with in the Baltic Sea.
280. Scymnus borealis Skoresby.—In the Arctic Ocean.
281. Pristiurus melanostomus Bup.   Ditto.
282. Raja clavata L.—In the Arctic Ocean, White Sea and Black Sea.
283. Trygon pastinaca L. — Everywhere in the Black Sea and from it enters the Sea of Azoff.
284. Petromyzon Wagneri Kessl.—In the Caspian, enters all its rivers.
285. Petromyzon fluviatilis L.—Basins of White Sea and the Baltic, with Lakes Ladoga and Onega.
286. Petromyzon Planeri Bloch.—In the small rivers of the Baltic, Black Sea and Caspian.
287. Myxine glutinosa L.—In the Arctic Ocean, White Sea and Baltic.
288. Branchiostoma lanceolatum Pall.—In the Black Sea.

It will be seen from this list, that 58,6°/o of the fish of European, Russia are inhabitants of salt or brackish water, 32,3°/o are fresh water

fish, and 8,6°/o are migratory. Of the number of sea and brackish water
fish (169) only 2—3°/o are available for trading purposes (according to
whether those fish are to be counted or not, which are caught by chance
or very seldom).

Of the fresh water fish (93) about 82,8°/o are caught. All the
migratory fish as fit for trading purposes. Of all the number of our fish,
about 44°/o are caught in the fisheries.

This shows of what great consequence the fresh water and migratory
fish are to us; this will be seen more clearly, when we enter into the
particulars of fishery statistics, which show that, by weight, salt-water fish
of form a very insignificant part of the whole amount caught.

III. A survey of the more important kinds of fish.

The Perch (*Perca fluviatilis L.*) is found in all the rivers and
lakes of European Russia, and also in the brackish water of the Caspian,
the Sea of Azoff and the Baltic.

The largest specimens have been caught in lakes, as for example,
in Lake Seliger, thery have reached 8 lbs, and in the Trans-Ural lakes
from 10 to 12 lbs, although the normal weight does not exceed 3 lbs.
It is caught from the length of an inch and three quarters, and goes
into the market in a fresh, frozen, partly in a salted, more often
in a dried state; this dried perch goes by name of *soosh* or *sooschtch*, of
which more will be said afterwards.

The Gremille (*Acerina cernua*) is lives everywhere in fresh water, with the exception of the rivers falling into the southern part of the
Caspian. It reaches its greatest size in the waters of Siberia, where specimens have been found 46 centimetres long and weighing $1^1/_2$ lbs. In
European Russia the weight does not generally exceed $^1/_5$ or $^1/_3$ of a
pound.

In the governments of Olonets, Novgorod, Tver and Pskoff, where
the lakes teem with perch and gremille, it is not so much the full-grown
fish that is caught, as chiefly young fish from $^7/_8$ to $3^1/_2$ inches long,
which is made into an original kind of preserves, called *soosh*.

These small fish are caught with floss-silk nets, and are then dried
in ovens; some 10 lbs of *soosh* is got from 40 lbs of raw fish.

The quantity of *soosh* prepare on the lakes Latch, Tchiranda, Koobinsk, Seliger, Peipus, Velie and the mass of smaller lakes, is in any case not less than 75,000 poods (40 ℔ Russ. to a pood). Therefore, some 300,000 poods of young fish must be caught, or about 4,000,000,000 sh, as from 12,000 to 80,000 go to a pood of *soosh*.

Soosh is sold at present for 10—11 copecs a pound.

The Soodak (*Lucioperca sandra*) is found everywhere, in the sea and in rivers, but is caught chiefly in the basin of the Caspian Sea, where, however, it is caught along with *Lucioperca marina Cuv.*, a sea variety, the fishermen making no distinction between them.

The largest specimens weighed 25 ℔, and were about 3 feet long. The average weight in whole-sale trade at present is about 5 lbs, while it used to be 9 lbs in former days. It is not so very long ago since 45,000,000 *soodaks* were exported from Astrakhan; in the fifties of this century Baer fixed the number at 28,500,000, which weighed 2,500,000 poods, at the present time, about 26,000,000 *soodaks* are exported from Astrakhan, weighing as much or 2,000,000 poods.

Besides this, some 7,000,000 *soodaks* are exported from the r. Kooban.

The *soodak* also yields e great quantity of roe, which is sold together with the roe of the bream and Caspian roach, under the name of *tchastikovi* (or scaled-fish) caviar; it is exported chiefly to Turkey and Greece. As much as 100,000 poods of *soodak* caviar used to be exported from Astrakhan, but of late years only 50,000 poods, owing to the diminution of the size of the fish caught. When carefully prepared, the oil of the *soodak* has a very fine taste.

The Silure (*Silurus glanis*) is found everywhere, except in the basin of the White Sea portion of the Arctic Ocean, but it is found chiefly · in the basin of the Black Sea and the Caspian. It weighs as much as 8 poods, and it *has* reached the great weight of 15—16 poods. In Astrakhan the average weight of the silurus met with in the markets is from $1^1/2$—2 poods.

As much as 100,000 poods of salted silurus are exported from Astrakhan; in 1881 64,000 of them were caught in the river Koora; some 20,000 poods are caught in the Black Sea.

The general amount caught may therefore be estimated at about 250,000 poods. As the whole-sale price of the silurus is from 40 cops. to 2 rbles per pood, the price of the whole haul must be about 250,000 rbles, not counting the price of the isinglass, some 1,000 poods of which is procured.   2*

The Salmon. There are two kinds of Salmon, which are found in the two different parts of Russia in Europe — *Salmo salar*, which is found in the basin of the White Sea part of the Arctic, and the Baltic basin, and the *Caspian salmon* (Salmo caspius Kessl.) which is found in the southern and central parts of the Caspian, from where it goes up the rivers to spawn, chiefly ascending the rivers Terek and Koora.

The average weight of a Caspian salmon is 20 lbs.

In the north some 50000 poods of salmon are caught yearly.

In the basin of the Baltic the yearly haul does not exceed 5000 poods.

In the Koora more than 30000 Caspian salmon are caught; as, for instance, from 1863 to 1868 204123 salmon were caught, making it about 34000 a year. In the last two years, owing to the abolition of the farming system, and to free fishing, much less salmon has been caught, viz. in 1881 — 12626 salmon, and in 1882 — 9978 salmon.

Besides there two species of salmon, in the north, the salmon trout and *Salmo salvelinus* are of some consequence in trade; and in the Caucasus, in Lake Goktcha — *Salmo gegarkuni* and *Salmo ischchan.*

The Smelt (*Osmerus eperlanus*) and its lake variety the sparling, *Osmerus spirinchus*, are found in the basin of the Baltic and of the White Sea, but the sparling has found its way into some of the lakes of the Volga river-system. It is sold fresh, frozen, and dried (with a little salt).

The price of the sparling, when sold wholesale by the fishermen, varies in different places and at different times of the year, from 30 copecs per pood (the spring sparling on the Talabsk islands on the Pskoff lake), to 2 roubles 20 copecs (the winter sparling in the smaller Novgorod lakes).

In the Pskoff and Tchoodskoë Lakes up to 100.000 tchetverts *) of sparling is caught, for the sum of 500.000 roubles. Lake Belozero yields 40.000 poods. The whole amount of smelt and sparling caught in our northern lakes is about 1,000.000 poods.

The Beloribitsa or White Salmon (*Luciotrutta (Coregonus) leucichthys*) is found in the Volga and its tributaries and in the Ural, and the Nelma (*Coregonus nelma*), which is very like it, is found in the Koobinskoë Lake, North Dvina and Petchora.

It is one of the best and most valuable fishes, with which none of the other coregoni can be compared as to taste, not even the famous Mucsun of Liberia.

---

*) a tchetvert is equal to 5,775 bushels.

The back of the white salmon is dried and goes by the name of *balik*.

At the beginning of this year (in February) this fish was sold in Astrakhan at the rate of 11 roubles 20 copecs per poods, and the *balik* at 20 copecs per pood.

About 30.000 poods of white salmon a year are exported from Astrakhan; while the whole amount of white salmon caught not only at Astrakhan, but also in the Ural, all along the Volga and Kama etc., and of the nelma caught in the basin of the Arctic and the White Sea very likely reaches 100.000 poods, for the sum of more than 1,200.000 roubles. I may here remark, that by D-r Oldekop's observations, the flesh of the white salmon is twice as nourishing as that of the carp.⁻

The S i g (*Coregonus*).

Russia is very rich in varieties of the *Coregonus*. The latest investigator M-r Polyakoff, has discovered 35 species in Russia; of these, 8 species are known in the basin of the Baltic (*Coregonus albula, C. Nilssonii, C. fera, C. Baerii, C. Widegreni, C. lavaretus, C. maraena, C. tscholmugensis*).

Of the siberian *coregoni*, brought froz-en to Petersburg, the largest and tastiest in the *Coregonus muscun*.

It is impossible to say even approximately what is the amount of sigs caught, but there is no doubt that the number is yearly decreasing in European Russia.

The C a r p (*Cyprinus carpio*). This central-asiatic fish is found in all parts of the Aral, Caspian, Sea of Azoff and Black Sea basins; it has reached the weight of 20—27 lbs as the length of about 3 feet. In the market the average weight is 9 lbs.

As much as 200.000 poods is exported from Astrakhan. In February (1883) at Astrakhan the carp was being sold at 80 cops per pood.

From 30.000 to 60.000 carp are caught in the Koora.

The R o a c h (*Leuciscus rutilus*) and its marine variety (*L. rutilus var. caspicus*).

This fish, notwithstanding its cheapness, is of great importance in Russia, being found in innumerable quantities and forming, in a dried state, a national dish.

The roach, living in the Caspian Sea and not going very for up the Volga, has formed a variety (*Cyprinus grislagine Pall.*), just as in the Black Sea the *Leuciscus Heckelii Nordm.*, called the *Taran* by the natives, is but a variety of the roach.

From 300,000.000 to 400,000.000 Caspian roach is caught in the Caspian Sea, which weigh (when dried) 3,000.000 poods, but as 1000 fresh roach weigh 18 poods, the weight of the whole lot will be equal to 5,400.000—7,200.000 poods.

How great was the demand for this fish will be seen from the fact that in 1868, in Astrakhan 1000 salt Caspian roach cost 60—80 copecs; and in 1872 1000 unsalted Caspian roach were bought at the fisheries for 7 roubles!

The roe of the Caspian roach, to the amount of several tens of thousands of poods, is sold together with that of the bream.

The Black Sea Taran (*Leuciscus Heckelii*) is caught in the Black Sea, and the rivers Kooban and Don — about 100,000.000 fish.

. The following fish, called at Astrakhan by the general name of *Taran* (which must not be mixed up with the *Taran* of the Black Sea, *Leuciscus Heckelii*), are also of great consequence in the fish-trade of the Caspian, and in the food of the poorer classes. These are:

The Blue Bream (*Abramis ballerus*), the Zope (*Abramis sopa*), and chiefly, the *Scardinius erythrophthalmus* and *Blicca björkna*. This *taran*, as also the razor-fish (*Pelecus cultratus*) when salted, is used in great quantities, chiefly by the peasants of Little Russia. 3,500.000 poods of it are said to have been exported from Astrakhan; but the real number must be about 300,000.000, which must weigh, when fresh, no less than 5,400.000 poods, counting 18 poods for 1000 fish. I will here remark that, according to D-r Oldekop the scales of the razor-fish can be used for making pearl-powder; and how great is the quantity of scales gathered?!

The Bream (*Abramis brama*) is found everywhere in Russia as far north as 63° n. lat.

The largest specimens have weighed 8 lbs, though the fishermen speak of 15 and 20 pound bream. The average weight of the bream sold in Astrakhan does not exceed 3,2 lbs, and in the north, for ex., in the government of Novgorod — $1^1/_2$ lb. About 400,000 poods of bream are exported from Astrakhan *). Besides this, about 150,000 poods of bream roe together with the roe of the Caspian roach, is exported.

In the north the quantity of bream has diminished to a considerable extent owing to the increased fishing, and the long time the bream takes to grow. Some fifty years ago in the government of Novgorod, in the small

---

*) According to M-r Kooznetsoff — 1,375.000 poods.

lake Velye, 50,000—60,000 2-pound bream were caught at one haul of the net; at the present time no more than 10—20 bream are caught, and very rarely are 5—10 poods caught.

The Bleak (*Alburnus chalcoides Güld.*) is found in great quantities in the Caspian, Sea of Azoff and Black Sea; it enters the r. Koora from the Caspian Sea.

The general amount of bleak caught in the Koora and the Sea of Azoff is said to be about 2,500.000.

Here again we may remark the scales of this fish, which could be used for pear-powder, are lost, just as many other things.

The Caspian Herring (*Alosa caspica*) and the Black Sea Herring (*Alosa pontica*).

Just as in Western Europe there is the *Alosa vulgaris* and *Alosa finta*, so in Russia, in the basins of the Black Sea and the Caspian there are multitudes of two independent species, closely allied to the European kinds, — the larger *Alosa caspica* and the smaller *Alosa pontica*. Both these species lives in the two above-mentioned basins, and, although they are distinguished by the fishermen, they are sold together.

It should be remarked that the so-called *poozanok* (*Alosa pontica*) though smaller in size, is much more tender and tasty than the larger, so-called *jheleznitsa*, and ought to be salted separately, like the Dutch herring; but our fish-dealers have not reached this, though, may be the time is at hand when they will turn their attention to this also, judging by the rapidity with which he herring-trade has developed here. Thirty years ago the herring was caught as the mouth of the Volga *) only for oil, and no one would eat it, thinking it to be a *mad* fish.

But since the expediton of M-r Baer, a member of the Academy, the Caspian herring began to be salted down,—in 1855 about 10,000,000 herrings were salted, in 1856 more than 20,000.000, in 1857 — more than 50,000,000, in 1871 as many as 140,000,000 were salted, in 1872—160,000,000, in 1873—180,000,000 and of late years in Astrakhan about 250,000,000 herrings were salted. It must be remembered that the Astrakhan herring is almost 3 times the size of the Duch herring, so that 250,000,000 herring, when salted, weigh 280,000,000 lbs!

The price of the Astrakhan herring is rising considerably; twenty years ago it was found profitable to procure some 30 lbs of oil from 1,000

---

*) The Russian name of this fish is *beshenka*, from *beshennii*, which means «mad».

herrings (costing about 60—70 cops) while at the present time the next catch (i. e. May 1883) has already been sold by some merchants at **12** roubles the thousand.

The Caspian herring is salted in boarded pits which can hold some 100,000 fish, and is then packed in barrels and half-barrels by **1,000** and **5,000**.

The Sea Herring (*Clupea harengus*) is caught in Russia in the White Sea and the Baltic.

The White Sea herring-fisheries are concentrated in the Gulf of Onega, where the herring come about the 23-d of April.

It is salted in kegs holding 20 lbs of fish.

The total is about 200,000 poods (according to the return of the Committee of Statistics). On the White Sea 3030 men, 1,659 large nets and 5,540 smaller tackle are employed in this trade.

The herrings salted by the Solovetsky Monastery are remarkable for their flavour.

In the Baltic the herring is very small and is known by a peculiar, Finish name—*salaka*. It is from 14—22 centimetres long.

I do not known how great is the quantity of herrings caught in the Russia part of the Baltic, but it must be very considerable, as the population of six governments (not counting Finland) feeds on this fish.

Baer brings an instance of 4—5,000,000 *salakas* being taken at one haul on the waters of one proprietor; also, April 4-th 1844, near the Woist estate 4,500,000 of these fish were caught by 12 or 13 nets; in 1838 5.200,000 *salakas* were caught on the Peddis and Majhin estates, and in 1844—4,800,000. There is no doubt that such hauls are still to be met with, as, judging by the data, from the end of the last century the hauls of *salaka* have not decreased, though they vary a great deal.

The Sturgeon species (*Acipensèri*).

The waters of Russia are very rich in these valuable fish, with which no other fish can be compared as to the flavour, nourishing qualities and the mass of useful products yielded by it the following species are found in Russia.

a) The Sterled (*Acipenser ruthenus*) is found in the rivers falling into the Black Sea and the Caspian, with the exception of the south Caspian, where solitary specimens are met with now and then (for instance, in the river Koora). The sterled is also found in great quantities in the rivers of Siberia, which fall into the Arctic Ocean, particularly in the Irtish. Then, some forty years ago the sterled penetrated through the canals into the North Dvina, and, finding the conditions favourable to its existance (for. ex. cold water, which is so necessary for it), it not only settled down and multiplied, but acquired some peculiarities in its exterior (a short, blunt snout and an arched back) and also a fine flavour, for which in Petersburg it is prized more than the Volga sterled. I must remark that even in the system of the Volga the sterled is much finer in the north (for instance in the r. Sheksna) than in the southern parts, and the further south one goes, the less tasty the sterled becomes.

The sterled reaches the length of a metre and the weight of 60 lbs. but such specimens are rarely met with; a twenty-pound sterled is no rarity, but the greater part of the sterleds caught and sold are generally from 30 to 50 centimetres long. Unfortunately, of late, on the Dnieper and even on the Volga they have begun to catch such small sterleds—only some 10—12 centimetres.

The Eastern Sturgeon (*Acipenser Güldenstädtii*) is found in the same places as the sterled, but keeps more to the mouths of rivers and goes further into the sea than the latter, for which reason it is found all over the Caspian, is caught in the Koora and the rivers of Persia; in the Siberian rivers it lives further north than the sterled.

The Russian sturgeon reaches the length of 270 centimetres and the weight of 200—240 lbs, though it has been asserted that there are larger ones, up to 520 lbs weight. But the average weight of the sturgeon sold in Astrakhan does not exceed 60 lbs, or the length of 180 centimetres.

The Western Sturgeon (*Acipenser sturio*) lives in the waters Baltic, and is rarely to be found in the Black Sea, where it finds its way along with a number of other animals from the Mediterranean.

The Sevruga (*Acipenser stellatus*) is found in the basins of the Black Sea and the Caspian, and does not go very far up the rivers. The fishing is concentrated chiefly on the r. Koora and the Ural. Although it reaches almost the same length as the sturgeon, the sevruga is much lighter; the average weight of an Astrakhan sevruga does not exceed 30 lbs, though a big sevruga weighs some 60 lbs. By Dr. Oldekop's investigations, it appears that the sevruga, of all the sturgeon kind, not excepting even the nelma, contains the greatest quantity of nourishment,—viz. to a pound of fish there is 53,6 zolotniks of nourishing matter, i. e. about 55,2%.

The Schyp (*Acipenser schypa*) is found in the same places as the sevruga, and besides that, in the Aral Sea, where there are no other sturgeon. It is rarely found in the Black Sea. The fishing is concentrated chiefly on the Koora, Ural, and Aral Sea. It weighs about 60 lbs.

The Big Sturgeon (*Acipenser huso*) or Beluga. This giant lives in the basins of the Black Sea and the Caspian. It attains a great size. For instance, in 1869 one was caught at Saratoff, which weighed 2760 lbs, in 1813 near Sarepta 3200 lbs., in 1847 in the Ural 1600 lbs. But it has been stated that such large belugas have heen caught, that the fishermen were not able to drag it out, and had to let it go, and that they must have weighed more than 4000 lbs. Of course, these are exceptionable cases, but a beluga of 800—1200 lbs. is caught pretty often; on the other hand, the minimum for the Astrakhan market is said to be 120 lbs, and the average, about 360—400 lbs.

(Besides the above-mentioned representatives of the genus Acipenser, we have two other species of the sturgeon in the r. Amoor and three species of the genus *Scaphirhynchus* in the rivers Amoo-Daria and Sir-Daria, which fall into the Aral Sea; these are *Scaphirhynchus Fedtschenkoi*, *Sc. Kaufmanni* and *Sc. Hermanni*; but they are of no consequence ins the fish-trade).

The sturgeon fish are sold either fresh (alive or frozen) or preserved, salted, dried in the sun (*balik*), smoked and pickled (the sterled). Besides this, the following valuable products are procured from them.—caviar, isinglas and *vyaziga* (dried spinal).

The roe forms, about $1/6$ of the weight of the whole fish; about $1/5$ in the sturgeon, $1/4$ in the *Schyp*, $1/10$ in the sevruga

How great the catch is, can be judged from the following figures —

According to Khlebnikoff from Astrakhan the export was:

Beluga	300,000	poods
Sturgeon, sevruga, schyp.	.1.000,000	»
Sturgeon roe	90,000	»
Isinglass from the beluga and sturgeon .	4,000	»
Vyaziga	4,000	»
Total, . .	1.398,000	poods.

The number of sturgeon caught in the r. Koora was:

from 1848 to 1854 on the average				562,715	fish
» 1863 » 1871 »	»	»		463,822	»
» 1872 » 1878 »	»	»		323,549	»
during 1881	»	»	»	455,593	»

According to M-r. Danilovsky, the amount af sturgeon caught yearly in European Russia is, in round numbers, 8.000,000 roubles' worth, viz.:

Isinglass to the amount of			600,000	rbles.
Vyaziga « »	»	»	100,000	»

Caviar to the amount of 2.250,000 rbles.

Flesh (balik etc)   ›   › 5.000,000   ›

L a m p r e y s. Of the three species of lamprey found in Russia, only two are caught. *Petromyzon fluviatilis*, which lives in the basins of the Baltic and White Sea, and *Petromyzon Wagneri*, which lives in the basin of the Caspian. The first kind is pickled; the second, which reaches the length of 70 centimetres, is caught in great quantities on the Volga and the Koora. It was not caught at all before 1870, being considered a snake; only on the Koora a small quantity used to be dried and used for lighting, instead of candles. From the year 1869 the lamprey began to be caught on the Volga, first at Saratoff, then lower down, as far as Astrakhan. In Tsaritsin and Tcherniy Yar a small portion of the lampreys caught is pickled in kegs and in this form sent to the markets, chiefly to St. Petersburg and Moscow; the greater part, about $^5/_6$ of the whole catch, goes for oil.

Here are the data, by which one can judge of the extent of lamprey-fishing; they were given me by one of the merchants. From Tsaritsin to Enotayevsk some 50.000,000 lampreys are caught weighing 150,000 — 175,000 poods, and yielding from 22,500 to 26,250 poods of oil. A pood of fish yield $5^1/_2$—6 lbs of oil. A thousand fresh lampreys weigh 3—$3^1/_2$ poods. The fishermen at Tcherniy Yar are paid 30 copecs per pood, and in Tsaritsin 50—60 cops, as there the lamprey is pickled, and only freshly-caught fish is taken. In Tsaritsin a pood (40 lbs) of lamprey-oil, along with the barrel, costs 3 rbles 50 cops.

IV. Fishing Statistics.

Some thirty years ago the Academician Baer proved by figures that the richest fisheries in the whole world were the Russian fisheries on the Caspian Sea; and yet twice as much fish is caught there now. But as yet there is no possibility of determining, with anything like accuracy, the amount caught in Russia, as, owing to the abundance of fish and the absence of supervision, statistics are taken in very few places, and they are further hindered by the custom of the traders to hide the amount of the hauls.

What statistics there are, are placed further on, according to the water systems; in many cases there statistics are more or less conjectural and are therefore only approximate. Nevertheless, I can boldly assert that the figures quoted are at least not higher than the real ones.

1. The Arctic Ocean (Mourman coast) and White Sea.

According to the statistics collected by M-r. Danilovsky, the whole fishing-products of the Arctic and White Sea basin (except Siberia) must be estimated at 1.000,000 roubles.

Further particulars can be given by the following numbers, communicated by M-r. Shoobinin.

On the Mourman coast cod is caught to the amount of					500,000	poods
halibut (*Hyppoglossus maximus*)	»	»	»	»	150,000	»
pikshooya		»	»	»	»	100,000	»
On the White Sea salmon (*Salmo salar*) is caught	»	»	· »	»	50,000	»
According to the Committee of Statistics, herrings are caught on the White Sea, at Soroki 200,000 poods, and in the other places 200,000 poods, total . .					400,000	»
Other fish is caught to the amount of					200,000	»

So that altogether 1,400.000 poods of fish are taken, which must cost at least the same number of roubles.

2) The Baltic Sea, according to M-r Soodakevitch's estimate, yields some 1,000.000 roubles worth of fish, counting the fish caught on Lake Peipus, where the sparling alone is 300.000 poods.

3) The Caspian Sea. There is no doubt that this is the richest basin not only in Russia, but in the whole world.

Thirty years ago, M-r Baer estimated the amount of fish caught in the Caspian at 11,000.000 poods, for the sum of 10,500.000 rbles, viz.

Sturgeon	2,200.000	poods
Soodak	2,500.000	»
Bream (28,500.000 on an average)	1,200.000	»
Caspian herring (100—160,000.000) . .	3,250.000	»
Silurus	185—200.000	»
Salmon	17— 40.000	»
Beloribitsa (Nelma) . . .	8— 60.000	»
Carp . . . . . about .	200.000	»
Pike	25 — 80.000	»
Caspian roach	250.000	»
Zherekh, bersh, roach, razor-fish,		

guster, zope (kind of bream),
blue bream, red–fin, perch,
gremille, carass . . . . 250.000 »
On an average . 11,000.000 poods.

But the circumstances have altered; a more thorough fishing has begun, and the hauls are 3 times as great.

According to the information given by fish-merchants, the following is exported from Astrakhan:

Caspian herring 250,000.000 fish		
(700.000 barrels) . . .	7,000.000	poods
Caspian roach 350,000.000 fish	3,000.000	»
Taran, zope, razor-fish and be-		
loglazka 200.000 fish . .	3,500.000	»
Soodak	2,000.000	»
Bream	400.000	»
Carp	200.000	»
Silurus	100.000	»
Beloribitsa (Nelma)	30.000	»
Beluga (sturgeon)	300.000	»
Sturgeon, sevruga, schyp . .	1,000.000	»
Black roe	90.000	»
Roe of the Soodak	50.000	»
Roe of the common and Caspian		
bream	150.000	»
Beluga and sturgeon isinglass .	4.000	»
Vyaziga	4.000	»

Total . 17,828.000 poods.

But these figures do not show the whole amount caught in the Caspian Sea and its tributaries, as this does not include the fish caught in the r. Koora (Caucasus), in the Ural as far as Orenburg, in the Volga above Astrakhan, and lastly, the fish used by the inhabitants, who live chiefly on fish. Besides, the weight shown is that of the prepared fish, while generally the haul is counted by the weight of the fresh fish.

Taking this into consideration, we must count that 6,500.000 poods of Caspian roach are taken, and 5,400.000 poods of *taran*. Then, by the latest intelligence, from M-r Kooznetzoff, 1.350,000 poods of bream are taken. Replacing the figures of the above tables by these figures, we get a total of 24,203.000 poods. Adding to this the fish that is not exported through Astrakhan, and that used by the inhabitants, we get the round sum of the fish caught in the Caspian (of course including the Ural, Koora and lower part of the Volga) which will scarcely be less than *28,000.000 poods*, and which certainly must be worth 35,000.000 rbls.

4) The S e a o f A z o f f, with the rivers Don and Kooban, yields, according to M-r Danilovsky's calculations, about ²/₃ of the amount caught in the Caspian Sea; therefore we can freely put it down for at least 6.000,000 poods.

5) The B l a c k S e a, with the lower parts of the rivers falling into it, i. e. the Dnieper, Bug and Dniester, yields very little fish, namely, only some 100.000 roubles.

6) The lakes and river yield a very considerable amount of fish. If we recollect that in our northern lakes the sparling alone amounts to 1,000.000 poods, that in spring much more than 100.000 sterled a year ore caught on the Volga, then, suppose, one cannot count the total of the fish caught on all our rivers and lakes to be less than 7,000.000 poods. So that all the fish caught in European Russia amounts to more than *40,000.000 poods.*

I s t h e q u a n t i t y o f f i s h i n t h e w a t e r s o f R u s s i a d e c r e a s i n g o r n o t?—This question has often been answered and, fortunately, in most cases in the negative.

There is no doubt about the quantity of sea-fish remaining the same. As regards other the cod, or the White Sea herring or the Baltic *salaka*, it cannot be said that they are decreasing.

As regards lakes and rivers and fresh-water and migratory fish the following can be said.

In our nothern lakes and in the upper (partly also in the middle) parts of the rivers fish has perceptibly decreased, and continues to de crease with the development of agriculture and trade, and with the in crease of population, which lead to the destruction of wild, over-grown river-banks. But in the lower parts of rivers, where, in consequence of natural conditions, agriculture cannot exist, and where large areas, cover ed with primeval verdure, present conditions favourable to the life of such a mass of insects, as Western Eurepe can have no idea of, the case is different, and no decrease of fish is to be noticed, except in a few cases. For instance, there is no doubt that the quantity of white salmon (*Lucio-trutta leucichthys*) has decreased in the Volga, as also the Caspian salmon (*Salmo caspius*) in the r. Koora, and various kinds of the genus *Salmo* and *Coregonus* in the Baltic. This decrease of such valuable and slow–breeding fish is owing to the way of catching them, by stretching nets across the whole breadth of the river.

The sturgeon species, owing to their great prolixity, have not decreased perceptibly in quantity, but they have decreased in size, in consequence of which the same weight of fish continues to yield smaller and smaller quantities of useful products.

In illustration of this I will bring forward some statistics, given by M-r D. N. Sokoloff, which refer to the fisheries on the river Koora.

The amount of sturgeon caught along the r. Koora below Salyan, and in the r. Akoosha.

In 1848	588,475 fish, which yielded 26,522 poods of roe
» 1849	543,019 » » » 30,095 » » »
» 1850	617,029 » » » 31,969 » » »
» 1851	520,987 » » » 28,484 » » »
» 1852	616,619 » » » 34.089 » » »
» 1853	571,641 » « » 31,784 » » »
» 1854	481,242 » » » 24,721 » » »

on an average 562,715 fish, which yielded 29,666 poods of roe

The average catch
from the year
1863 to 1771 463,822 »   »   »   24,596   »   »   »

The average catch
from the year
1872 — 1878 323,549 »   »   »   12,105   »   »   »

For 1881 . . . 455,593 »   »   »   10,145   »   »   »

«In the fifties and sixties of this century», says Sokoloff, «the roe of the sturgeon caught was about $\frac{1}{10}$ of the whole weight of the fish; in the seventies, in the lower and more abundant parts of the r. Koora the roe yielded by the sturgeon was about $\frac{1}{27}$ of the whole weight, and in the last year and a half (1880 and 1881) the roe of the fish caught in the waters under the supervision of the Fisheries Superintendence was only $\frac{1}{44}$ of the whole weight.

This considerable decrease of the quantity of roe yielded by the sturgeon clearly shows that much smaller fish is caught.

«By the words of old fishermen», says Sokoloff, «the average weight in the sixties was 400 lbs of roe for a beluga, for the sturgeon and schyp—40 lbs, sevruga—15 lbs; while at the present time they have grown much smaller:—an average sturgeon yields 25 lbs, and a sevruga no more than 10 lbs. This is also remarked everywhere in connection with fresh-water fish. Even though very large fish are still to be found, we never see such giants as there were before».

A few years ago, near Lake Onega, several article were discovered, which belonged to men who lived in the age of stone, and among these articles the bones of several fish (pike, silurus) were discovered, the size of which convinced us that in those distant days men caught pike $2^1/_2$ metres long in the same lake where now the pike is rarely bigger than 70 centimetres.

V. The price of fishing produce.

Of course, the prices of fish and of fishing-produce differ in the highest degree, and as the transport of fish, owing to the great distances between centres of trade in Russia, comes very expensive, the price of fish in places far off from the fisheries is tremendous compared with the price of fish at the fisheries themselves. It will be sufficient to state that in St. Petersburg a live *talmen* (or salmon-trout), weighing 4—5 lbs is sometimes sold for 25 rbles, and a sterled from the Dvina, 10 lbs weight, costs 60 rbles, 15 lbs—120 rbles, to give a notion of the high prices that some fish fetch here in St. Petersburg.

Of course, the prices are quite different in the centres of the fishing-trade. Here, on the contrary, every one is struck by the cheapness of fish. As for instance, on Lake Peipus the sparling is sold by the fishermen at 30—60 copecs per pood; good-sized perch costs 2 rbles per pood in the government of Novgorod. In December 1882 on the Black Sea herrings were sold at 3 rbles per thousand, etc.

Not being able to give an account of the price of fish all over Russia, I think it will be as well to give some information as to the price of fish and fish-produce in the Astrakhan Exchange in 1882.

The information is received from the Astrakhan broker, M-r Lesnikoff.

	1882. Per pood	
	from	to
Fresh fish.	R. C.	R. C.
Beluga	3 80	4 35
Sturgeon (full size—$3^1/_2$ feet & more). .	5 —	5 50
,, (half size—$2^1/_3$ feet)	4 —	4 25
Sevruga (full size—$2^1/_3$ feet & more). .	4 —	4 50
,, (half size—not more than $2^1/_2$ f.)	3 30	3 80

Sterled (full size—14 inches & more) .	2 —	2 50
,, (half size—less than 14 inches) .	1 —	1 50
Beloribitsa (white Salmon)	7 —	12 —
Carp	— 75	1 —
Soodak . . . . . , ...	— 75	1 30
Pike	— 75	1 30
Bream	— 60	1 —
Different small kinds of fish	— 30	— 60

Newly-salted fish.

Carp	1 20	1 60
Soodak	1 30	2 20
Pike	1 10	2 —
Cingel	— 75	1 —
Bream	1 20	1 60
Different small kinds of fish	— 50	1 —

Salt.

Beluga	4 50	5 --
full size sturgen	5 50	6 50
half ,, ,, ...	4 50	5 50
full size sevruga	4 50	5 50
hàlf ,, ,, ...	3 50	4 50
full size sterled	3 —	3 50
half ,, ,, ...	1 50	1 75
full size silure	2 50	3 —
half ,, ,, ...	1 50	2 —
full size carp	1 20	1 60
half ,, ,, ...	— 60	— 80
Herring (shad) full size per thousand . .	17 —	19 —
,, ,, half ,, ,, ,, . .	5 —	6 —

Salt fish, dried.

Carp	2 50	3 —
Soodak	2 40	2 90
Pike	2 20	2 70
Cingel	1 —	1 25
Bream, full size, per thousand	40 —	60 —
,, half size, ,, ,, . . .	20 —	30 —
Caspian roach per thousand (i. e. 25 poods) .	3 —	10 —
Different kinds of small fish	— 70	1 20
Beluga & sturgeon *balik*, ...	12 —	16 —

Beloribitsa (white salmon) . . . . . . 10 —   12 —
Caviar (large-grained). . . . . 30 —   60 —
„   (pressed) . . . . 25 —   75 —
Beluga isinglass . . . . 140 —   150 —
Stergeon & sevruga isinglass. . . . . 170 —   175 —
Sterled   ditto . . . . . . 50 —   60 —
Silure   ditto . . . . . 24 —   26 —
Vyaziga (dried spinal of the sturgeon) . . 40 —   45 —
Fish-oil . . . . 2 80   3 —
Seal blubber . . . . . : . . . 3 50   4 —

For comparison I will quote the prices in Astrakhan 15 years ago, in 1867.

		Per pood			
		R.	C.	P.	C.
Beluga	from	2	10	to	2	15
Sturgeon, full size	„	2	25	»	2	90
„ half size	„	2	10	»	2	15
Sevruga, full size	„	1	85	»	2	40
„ half size	„	1	30	»	1	55
Silure, full size			1	60
„ half size			—	90
Beloribitsa (white salmon)			2	—
Sterled			1	25
Carp			1	10
Large-grained beluga caviar . . .	from	16	—	to	25	—
„ sturgeon caviar . .	„	12	75	»	16	—
Pressed caviar	„	12	50	»	18	—
„ (lower sort) . . .	„	3	50	»	—	—
		Per thousand.			
Caspian herring	from	10	75	to	—	—
Bream, full size	„	24	—	»	27	—
„ half size	„	12	—	»	13	50

On the Mourman Coast cod is sold at 60 cop. per pood.

In Archangel salt cod is sold at 1 rble. to 1 rble. 50 cop. per pood, and small- size 70 cop. per pood.

Unclarified cod-liver oil costs 1 rble. 50 cop. per pood on the Mourman coast.

Spring salmon (*Salmo salar*) is sold in Archangel at 5 rbles per pood, and autumn at 10—15 rbles.

In Archangel, herrings from the Gulf of Kandalak are sold in spring for 30—40 cops the keg (weighing about 20 lbs), and in autumn 1 rble to 1 rble 30 cops.

VI. The preserving of fish.

The profit of preparing conserves depends only on the abundance of fish, but also on the cheapness of the preserving material, and also on the conditions of the fishing, the length of the fishing, the temperature etc.

Some experiments were made of preserving our more valuable kinds of fish in tins, and though these experiments were not crowned with success, this may be ascribed to the inexperience and want of energy, of the experimenters, the absence of sufficient capital; what is even of more consequence, the expense of the preserving material (for inst. oil) and the tins. On the other hand, preserving fish in the materials which are found in abundance in Russia has given very good results, and our caviar and balik are quite as good as any other preserves.

The cheapest, most natural and the most employed means of preserving fish are winter frost, summer heat, and salt. It is evident that where the winters are very often 8 months long, and the average temperature from November to February is 6° Celsius, as in Novgorod, 10,4°, as in Archangel or even 16° C. as in Tomsk, the most natural way of preserving fish consists in freezing it, and one finds frozen Siberian *mucsun* sold all winter in Petersburg.

Even in the south, where the winter is shorter, and the average temperature of the same months is not lower than 3,5° C., as in Astrakhan, fish is preserved by freezing (sturgeon, soodak). This is done by putting them into a sort of pond, unconnected with the Volga or barred from it, and when winter sets in, the fish is taken out with a landing net, frozen, covered up with snow and sent off for thousands of versts.

On the other hand, the absence of rain, and the heat during the summer months enable the fisheries to profit by the sun's rays for preparing balik and dried fish. The mean temperature of Astrakhan, it must be remembered, from May to September is equal to +21,5° C. and in Trans-Caucasia, on the r. Koora, rises 1° (in Baku +22,5° C.).

In the north, where the sun is of no use, the same result is ob-

3*

tained by means of wood, and stoves built for the purpose, used for preparing the cheapest fish-conserves (*soosh*), which satisfy the frugal taste of the northern peasant, with whom this *soosh* takes the place of the flesh of warm-blooded animals.

The third, and chief means of preserving is common salt, a large supply of which is found just in the very places which abound with fish, in the salt lakes near the Caspian and Aral Sea.

When in the post-pliocenic period the waters of the Caspian fell, many lakes were formed, the waters of which evaporating, left great tracts of salt. This is still to be seen in many lakes, the greatest of which is Karaboogaz.

This is an enormous gulf, connected with the sea by a narrow and shallow channel, in which there is a constant swift current from the sea into the gulf, bringing daily into Karaboogaz 4.800,000 cubic fathoms of water (according to Baer, but Korskin counts 18,000,000) and with it 32,073,600 poods of salt, as the Caspian contains $1,3\%$ of salt ($1,32784\%$ according to the analysis made of the water in the Bay of Baku); as the water of Karaboogaz is evaporated by the rays of the sun, which heat the sand of the surrounding desert to a fearful degree, and as there is no fresh-water tributary, this quantity of salt settles to the bottom in the form of a thick crust, over which is a strong solution of salt, containing 284,996 pro mille (Rb. $Cl—0,251$, $KCl—9,956$, $Na\ Cl—83,284$, $MgCl_2—129,377$, $MgBr_2—0,193$, $MgSO_4—61,935 = 284,996+H_2O—715,004=1000.00$). «There is here» says prof. Schmidt, who analyzed some samples of Karaboogaz water sent him in 1876, «an enormous supply of Glauber's salts, which would furnish all the glass-work for a century»).

Of all the salt lakes despersed over the governments of Astrakhan, the largest are Elton and Baskoontchak. The first covers an area of 56,000 acres, and the latter 31,700 acres, with a supply of salt of about 907,000,000 tons.

Besides this, in Tchiptchatch there is rock-salt of a very good quality to the amount of about 170,000,000 tons.

The following tables, taken from an article by M-r Kooznetsoff, will show the amount of salt got in the government of Astrakhan, and the amount used there for salting fish.

The amount of salt got in the government of Astrakhan (in tons).

	1875	1876	1877	1878	1879	1880	1881	1882
inBaskoontchak	91881	106366	113554	133068	167231	168543	208215	213890
in Elton	8112	3260	3801	5971	1704	9231	28188	40406
Tchaptchatch. .	28282	16953	2558	20815	19616	15421	6556	8238
in the small Astrakhan lakes	31440	47906	53913	83885	84687	75986	133542	180704
Total . .	154715	174485	173826	243239	273238	269181	376501	443238

The amount of salt used in the government of Astrakhan (*chiefly for salting fish*) in tons.

1875	1876	1877	1878	1879	1880	1881	1882
64823	61905	53937	79955	—	—	76488	54809

VII. The exportation and importation of fishing-produce.

However great the amount of fish in Russia may be, it is still not enough for a population of nearly 100.000,000 with which, to a very considerable extent, fish takes the place of meat. Therefore we import from Europe, in round numbers, as much as 5.000,000 poods of fish yearly, for the sum of 7.000,000 roubles.

During the last 5 years, i. e. from 1877 till 1881 inclusive, the following was imported from abroad.

1. pickled fish, fish prepared in oil, in tins etc.

		On an average.	
and caviar:			
poods.	rbles.	poods.	rbles.
273,157 for the sum of 4.236,043		54131	847,208

2. salt and smoked fish:

20,758 for the sum of 275,983		4,151	55,196

3. Cod and dried cod's heads:

3.568,479 for the sum of 2.908,074		713,696	581,615

4. Smoked herring:

1,222 for the sum of 5,977		244	1,195

5. Salt herring in large barrels:

19.470,330 for the sum of 25.198,378		3.894,066	5.037,875

6. Salt herrings in small barrels:

251,317 for the sum of	1.097,157	50,263	219,431

7. Norwegian herrings:

47,336 for the sum of	29.759	9,467	5,952

Total: 23.632,599 for the sum of 33.742,371   4.726,520   6.748,474

Comparing these figures with the amount of fish imported in former years, for instance, from 1872—1876, we see that there is no difference to speak of in the amount of 1, 2, 3 and 4; as regard № 6 (herrings in small barrels) a slight increase has been remarked; as regards № 5 and 7 a decrease, viz. in these five years the following was imported:

5. Salt herrings in small barrels: On an average.

poods.		rbles.	poods.	rbles.
21.200,750 for the sum of	23.641,557		4.240,150	4.728,311

7. Norwegian herrings:

144,307 for the sum of		83,112	28,861	16,622

Therefore in these years, we imported some 365,478 poods less of these herrings per annum than during the time when the Caspian herring was not salted in such quantities in Astrakhan. This difference will be even still greater if we compare the importation of very early times with the present.

The exportation of fishing-produce is not great, and differs very much from the importation in its value. We export, in round numbers, only some 300,000 poods, but for the sum of 2.800,000 rbles. Therefore, paying 1,4 rbles per pood of imported fish, we take 9,3 rbles. for a pood of the fishing-produce we export; this is, of course, in consequence of the great value of the caviar and isinglass exported, and the low price of the herrings in large barrels and of Norwegian herrings.

For the 5 years, 1877 —1881, the following was exported from Russia Russia into Europe:

1. Sturgeon caviar: On an average.

poods.		rbles.	poods.	rbles.
536,593 for the sum of		1.697,063	107,318	339,412

2. Black caviar:

207,067 for the sum of		7.233,470	41,413	1.446,694

3. Different kinds of fish:

566,689 for the sum of		1.756,777	113,338	351,333

4. Isinglass (sturgeon, beluga):

12,229 for the sum of		983,290	2,446	196,660

5. Isinglass (other kinds of fish)
    45,015 for the sum of   869,126   9,003   173,825
6. fish oil:
    34,737 for the sum of   187,603   6,947   37,520

Total: 1.402,330 for the sum of 12.727,329   280,466 2.545,466

On comparing it with the first five year (1872—1876) it appears that the exportation only of №№ 3 and 5 has increased, viz.

3. different kinds of fish:
    poods.   rbles.   poods.   rbles.
    290,684 for the sum of 1.008,139   158,137   201,628

5. Isinglass (of the commoner kinds of fish):
    28,837 for the sum of   889,873   15,767   177,974

I may here remark that the importation of herrings must, in the near future, decrease considerably; the r. Ural can give as many as 100.000,000 Caspian herring, but it was not fished for until last year. The decrease of the sturgeon will, of course, cause the Caspian herring to be more sought after.

VIII. Character of the progress of fishing in Russia.

It is not 50 years since all our fisheries were almost exclusively occupied with the more valuable kinds of fish—sturgeon, salmon and the sig. At that time, for instance, no one on the Volga would think of eating the caspian herring (*Alosa*), considering it a mad fish, and therefore unwholesome. Some fifteen years ago, lampreys were not caught. The razor-fish was also not used for food for a long time, and even in 1850 and 1860 it was used for oil, as well as the taran and the Caspian herring, just as the lamprey is now caught for extracting oil.

Even the more valuable Kinds of fish, as the soodak (*Lucioperca sandra*) went for oil, notwithstanding that it contains so little of it.

But years went by. The population increased. Railways were built, which made it possible to bring fish-produce at a cheap and rapid rate to places not connected with the great fishing centres by water-routes. In consequence of this, the demand for fish increased, and of course the sterled and nelma (*Luciotrutta leucichthys*) could not satisfy it.—Then not only a more thorough fishing commenced, but many less valuable fish were salted and pickled, ond began to be used more seldom for extracting oil. The Caspian herring, razor-fish and taran not only became favourite dishes of a great population, but the money made by the sale of these fish was about twice as much as that made by the more expensive kinds of fish. We see that even the Volga lamprey is not only caught and oil extracted from it, but it is even pickled, and ten years will not pass be-- fore it will be found unprofitable to use it for oil. But this does not mean that our fisheries work as thoroughly as they could.

There are a great many very fine fish (as far as one can judge) which have not yet been caught regularly. For instance, there are millions of sardines (*Clupea delicatula*) and small herrings (*Clupeonella Grimmi*) which are never caught, though they come up the Volga in innumerable multitudes; as well as the sturgeon (*Altherina pontica*). The chabot (*Cottus gobio*), though there are great quantities of it in the Black Sea and the Caspian, is caught very seldom, and only with a fishing-rod. The Caspian herring, so much prized in Astrakhan, is never caught in the Ural, although there are millions of it there.

In the north, in the basin of the Baltic, the chabot (*Cottus gobio*) and the stickleback (*Gasterosteus*) might be used for oil, but they are not caught, and do much damage to the spawn of the salmon and the sig (*Coregonus*). But even the more valuable fish are partly wasted; many parts are thrown away at the fisheries, although these parts might easily be employed for some useful purpose. I have already pointed out how the scales of the shemka, the razor-fish and the uklei (*Leuciscus*, a sort of blay) might be used for making pearl-powder; the scales of most fish can be used for making glue. There are regular mounds of these scales lying useless at the Astrakhan fisheries, waiting for an enterprising person, who, by applying some technical knowledge, together with a little capital, will easily make his fortune by them.

IX. Pisciculture.

In Russia pisciculture began some thirty years ago. In 1855 it was simult aneously started by two persons: in Tagil (on the Ural) by Malysheff, a medical apprentice, by the orders of the proprietor of the Tagil works, M-r Demidoff, and in the Deman district of the government of Novgorod, by M-r V. P. Vrassky, a landowner.

Malysheff tried his experiments on the *Gadus lota* and with full success. Vrassky, having studied artificial fertilization on the spawn of the salmon, river-trout, the sig, perch and gremille (*Acerina*) etc., built a piscicultural establishment on his estate Nicolskoë, after the pattern of the Güchelgen establishment in Alsace, but with such conveniences, that to the present day it is working without any change.

Among other inventions and improvements in pisciculture, we owe Vrassky the discovery of the best way of fertilizing the spawn of the fish— that which is at present used everywhere, and is known under the name of the dry, or Russian fertilization.

After the death of Vrassky (1869), the Nicolskoë establishment was bought by the Government, and, coming under the jurisdiction of the Ministry of State Property, committed to the charge of the Inspector of Agriculture, M. K. Repinsky.

In 1871 Repinsky began selling (at Nicolskoë) the fertilized spawn of the sig and trout, and also the young, artificially bred, fish of these species, for the transmission of which he invented a very simple apparatus, which has alway given full satisfaction, and is employed in Russia to the present day.

After Repinsky's death, the Nicolskoë Piscicultural Establishment came under the management of Dr. Grimm.

In 1881, a branch of the Nicolskoë Piscicultural Establishment was started at St. Petersburg, in the Imperial Agricultural Museum, chiefly for breeding salmon and talmen (*salmo fluviatilis*).

In 1882, a small piscicultural establishment was started at Samara by the help of a M-r. P. N. Tichinsky, for breeding sterled and other kinds of sturgeon; also for breeding the Nelma (*Luciotrutta leucichthys*) and carp.

Besides breeding fish, the Nicolskoë establishment, having such a quantity of fish in its lakes and ponds, tries to solve all kinds of scientifically useful questions as regards fish and pisciculture. During the last four years (and partly even now) the following questions have been investigated:

1. The cross-breeding of species of salmon with the sig.

2. The causes which produce the development of hybrids.

3. The spawning time of fish living in the lakes near the Nicolskoë Establishment.

4. The mutual relations of the different inhabitants of the water basins.

5. The fauna of the lakes in the government of Novgorod.

Besides this, the Nicolskoë Establishment receives all those who wish to study pisciculture, and lately there were several students at the Nicolskoë Establishment who pursued their studies with great success, and who, at present, are managers of several private piscicultural establishments.

A short report of the work done by the Nicolskoë Establishment is published annually in the Journal of the Ministry of State Property.

But to return to the development of pisciculture in Russia. We must remark that the failures in pisciculture in western Europe have affected us, and caused a general distrust, which was increased by the death of our original pisciculturist, Vrassky, a man who firmly believed in the future success of pisciculture and who showed so much energy in the pursuit of his aims.

At that time our pisciculture was restricted, one may say, to scientific experiments on artificially fertilizing the spawn of the sturgeon. These experiments were fully succesful.

In 1869 the academician Ovsiannikoff ascertained the time for the spawning of the sterled and sturgeon, and tried, with success, some experiments ot fertilizing their spawn. He brought several thousands of young fish (reared by him) to Moscow and to St. Petersburg. From this time the culture of the sturgeon species became not only possible, but in many ways very convenient, and was tried many times by various persons, among whom we may notice E. D. Peltzen, who fertilizes some tens of thousands of sterled spawn yearly.

In the last few years, the time for the development of pisciculture in Russia has evidently arrived; doubtless, this is partly owing to the brilliant results attained in America. This development has shown itself in the increase of the number of persons interested in pisciculture and in the foundation of several private piscicultural establishments.

The following private piscicultural establishments have been added to the two already in existence (i. e. Senator Zeimern's near Tsarskoë Selo and M-r. Sakharoff's in the district of Jamburg):

1. R. A. Mooshinsky's in St. Petersburg.

2. W. L. Greig's in Kurland.

3. One in the Solovetsky Monastery on the White Sea, not to speak of some small fish-ponds (piscinia) for hatching fish from spawn bought at the Nicolskoë Establishment.

1. One attached to the district court of Pskoff.

2. P. A. Vasiltchikoff's in the Serpookhoff district of the government of Moscow.

3. In the Moscow Zoological Gardens and other places.

The Astrakhan Committee of Fisheries resolved (in 1881) to build a great piscicultural establishment at Astrakhan, for breeding sturgeon and nelma (*Luciotrutta leucichthys*) for which purpose 70,000 roubles were assigned.

Of course, Russian pisciculture is still in a primitive condition, and is far from being carried to the degree it might be. But, indeed, it cannot be otherwise, as our agriculture has not yet been able to get into its proper groove after the liberation of the serfs, and therefore, it is in a transition period, which will doubtless lead to a better future for all things, pisciculture among others.

X. Hunting and trapping in the waters of Russia.

The Russian seas, and partly the lakes, as pretty well stocked with animals that are useful for commercial purposes. The trade is not very much developed, as the greater part of the Arctic coast is seldom, if ever, visited by trappers and traders.

The following animals are hunted and trapped.

BEASTS OF PREY.

The Polar Bear (*Ursus maritimus*), which is found on the islands and icebergs of the Arctic Ocean. The extent of the trade is no known, but must be considerable.

The Sea Otter (or Kamtchatkan beaver, *Enhydris Stelleri*) has a very valuable fur, known in the fur-trade as the Kamtchatkan beaver. At one time it was very common in Kamtchatka and on all the islands between the latter and America; between 1743—1797 about 2111 furs were yearly procured in 1798—1822. The American Company got some 3600 furs yearly, and in 1822—1841 1161 furs. From that time the number of otters killed decreased steadily, and at present it has reached a very low figure; what is more, now it is not to be found at all in Kamtchatka, and is killed only on the Medniy (or Copper) Island.

Thanks to the kindness of M-r N. A. Grebnitsky, the manager of the Commandor Islands, I am enabled to give the following table of the number of otters taken on the island Medniy from the year 1872—1882.

In 1872 . . .   9 otters were taken
» 1873 . . . 14 »   »   »
» 1874 . . . . 54 »   »   ı
» 1875 . . . . 48 »   »   »
» 1876 . . . . 33 ·   »   »
» 1877 . . . . 68 »   »   »
» 1878 . . . . 94 »   »   »
» 1879 . . . . 2 »   »   » (cast out)
» 1880 . . . . 182 »   »   »
» 1881 . . . . 190 »   ·   ıı
» 1882 . . . . 115 »   »   »

in 11 years . . 809 otters were taken.

The average is therefore $73^1/_2$ a year.

The common or river otter (*Lutra vulgaris*) is found everywhere, but is rarely killed, and then only by the way.

RODENTS.

The River Beaver (*castor fiber*) was formerly spread almost over all Russia, and in considerable numbers; but it is dying out, so that now it is nowhere a great article of trade. At the present day there are certain places in European Russia where beavers are still to be met with, viz. In the government of Kiev, Poltava, Podolsk and Minsk; on the Niemen, in Lithuania, on the rivers Viatka, Petchora and N. Dwina; in the Ural and in the Caucasus (r. Aladan). In Siberia the beaver is still to be found in great numbers, and in some places is hunted, viz on the r. Pelish. But even here they are fast dying out during this century.

PINNIPEDIA.

The Seal (*Phoca annelata*).
The Sea-calf (*Phoca vitulina*).
The Kamtchadal or bearded seal (*Phoca barbata*).
The Greenland seal (*Phoca Groenlandica*).

These four kinds of seal are found in the Arctic Ocean and the White Sea and are the objects of a pretty considerable trade which is chiefly carried on by the inhabitants of the Mesen and Kem districts, and partly by those of the Onega and Archangel disticts. Also by the Solovetsky Monastery. In 1882 about 4164 people were in the seal-fisheries.

Seals are caught in very considerable quanties, and in a good year some seventy or eighty thousand poods of blubber have been extracted; in bad years about twenty or thirty thousand. In 1882, according to the newspapers, 9544 seal-skins and 32,485 poods of blubber were taken, and the money made by the blubber was about 55,890 roubles. But as the walruses and white whales (*Beluga Castodon*) were counted in, the last year was a bad one as regards seal.

The Tevyak (*Cystophora cristota*) is found along the Mourman Coast.

Of the above mentioned species *Phoca vitulina* is still found in the Baltic and lake Ladoga, where it is generally shot. The number of seals shot in the Baltic is not known, but in lake Ladoga about a hundred a year are killed, and during a very good year, a thousand, which give about 650 poods of blubber.

The number of seals killed on the Baikal is unknown, as also that of the seals killed in Siberia along the Arctic coast.

The Caspian seal (*Phoca caspica*) is found in all parts of the Caspian *), but chiefly inhabits the north-cast corner, where the sea is covered with ice in winter, and, what is of greater consequence, more of the smaller kinds of fish are to be found, on which the seal feeds. However, seals are also killed along the south-west coast of the Caspian, where there is plenty of fish.

The chief hunting-ground is the island of Soolsma.

Seals whelp about the 10-th of January; and for this reason it is forbidden to kill seals from the 15-th of December till the 15-th of February.

The Caspian seal is from 3 to 6 feet long, weighs about 2—4 poods and gives some 27 ebs of blubber.

Like all other hunting and trapping, seal-hunting in the Caspian is subject to sonsiderable fluctuations. In 1871, about 79,442 seals were taken, and in 1872—159,479. Therefore one must judge by the average of a considerable length of time. According to Sokoloff, from the year 1867 till 1873, about 958,959 seals were brought to Astrakhan, which makes the average of 136,994 seals a year. During the same time 634,025 poods 30 pounds of seal-blubber were exported, making it about 92,008 yearly.

---

*) The notion that seals are fond in the Aral is as incorrect as M. W. Wagner's statement that there are dolphins in the Caspian.

The Sea - bear (*Otaria ursina*). When we ceded our American possesions, we retained the Commandor Isles; these islands are the breeding places of the sea-bear, and here the hunting takes place. An American company has rented the hunting ground from the Russian Government for years. Till 1877, the Company used to pay 2 roubles for every fur taken; but since 1877 it has been paying 1 rble 75 cops, for the first 30,000 furs. The Company took, from the year 1871—1882 *).

In 1871 . . 3,412 furs were taken, and duty paid ?
» 1872 . . 29,318 » » » » » » 58636
» 1873 . . 30,396 » » » » » » 60,792
» 1874 . . 31,272 » » » » » » 62,544
» 1875 . . 36,274 » » » » » » 72,548
» 1876 . . 26,960 » » » » » » 53,920
» 1877 . . 21,533 » » » » » » 38,224$^{1}/_2$
» 1878 . . 31,340 » » » » » » 54,845
» 1879 . . 42,752 » » » » » » 78,004
» 1880 . . 48,504 » » » » » » 89,508
» 1881 . . 43,522 » » » » » » ?
» 1882 . . 45,620 » » » » » » ?

Total 390,903 seals during 12 years, and in the last 6 years, more than 233,271; while in the first 6 years, only 157,632. Therefore there most have been 75,639 seals more. But these figures are scarcely likeey to show the real number of seals killed on the Commandor and Seal islands. It is very likely that a good many sea bears are killed by poachers; and to judge by the words of a competent person, that «the rumour is persistently circulated in the London fur-market, that in Japan there are 20,000 furs taken by poachers on the Company's grounds», in consequence of which the price of the fur is the sea-bear immediatly fell $23^{1}/_2°/_0$, the number of sea-bears killed this year by the Company does not exceed 70°/₀ of the whole number killed.

The Company gets, as far as is known, an income of about 400,000 rbles from these sea-bear hunting-ground.

The Walrus (*Thrichechus Rosmarus*) is found in Nova Sembla, Vaigatch, and Kalguska, where the hunters go from the coasts of the White Sea. The walrus weighs about 100 poods. From 10 to 15 poods and even as much as 28 poods of raw blubber are got from each. When

---

*) This information is from M-r. N. L. Grebnitsky.

boiled down, the blubber yield about $^2/_3$ of oil, therefore one can get from 7—10 or even 18 poods. A whole, uncut skin costs 10—12 rbles. When cut into strips, it brings some 40 rbles.; the two tusks cost about 12 rbles. Therefore, a large walrus will bring in about 69—97 rbles, and an ordinary one not more than 45 rbles. Generally each vessel does not get more than 10 walruses, although there have been cases of 50 being taken. The average number of walruses taken in Russia is about 600, about 30,000 roubles worth. But this does not include the warluses killed by the aborogines of Siberia.

THE CETACEA.

The Beluga, or White Whale (*Delphinapterus leucas*), is from 14—25 feet long. Beluga-fishing is carried on in the White Sea, where the beluga lives all the year round; also in the gulfs of the N. Dvina, Onega, Kondolon and Mezen; in the Arctic Ocean it is found to the east of the White Sea, near the mouth of the Petchora, along the Timan coast, chiefly near the r. Piosha; near Nova Zembla at the mouth of the Obi and further on. In chasing fish, it goes very high up the rivers, for instance, up the Obi. It is caught with nets, with which it is surrounded, drawn to a shallow place and killed in what is called the dvor, or yard; from four to six boats take part in the work.

The quantity of oil got from the beluga is various. Sometimes a herd of large animals have been killed, each of which yielded about 12 poods of blubber, and at other times one meets with belugas that yield only some 4—5 poods.

The exact number of beluga caught in a year is not known, as in the statistics of the fisheries, the beluga is classed with all the walrus, seals, whales, etc.

The Dolphin (*Delphinus delphis* and *Delphinus phocaena*) is found in considerable numbers in the Black Sea. From it, in chasing fish, it enters the various gulfs and bays and into the Sea of Asoff. The Turks come into the Black Sea after the Dolphin, chiefly visiting Pischoonda. Our fishermen sometimes catch it, but generally content themselves with a stray dolphin that may get in among the fish. *Delphinus phocaena* is sometimes met with in the Baltic, and even has come up as far as Cronstadt (but very rarely).

There are four kinds of whales in the Arctic Ocean.

Megaptera boops.
Balaenoptera laticeps.

Balaenoptera musculus.
Balaenoptera Sibbaldii.

The last kind is the one that whalers chiefly kill, the three first being killed now and then.

Notwithstanding the efforts of the Russian Government to increase whaling, it is still in a very primitive condition here. The Laps and Pomors, it is true, use whale-blubber, but it is procured from the carcases of whales that are often driven ashore. They never kill whales, owing, perhaps, to the false idea that the whale drives the fish *Moyva (Mallotus arcticus)* to the shore, and that therefore, whales are useful to the fisheries, and that they ought not to be exterminated. However, from 150 to 200 whales a year are killed on the Mourman coast by Norwegian whalers, who have oil-works in Finmarken. How profitable whaling is, will be seen from the fact that all the expenses of the trade are covered by the sale of the secondary products, such as whalebone etc., and that the oil, of which each whale yields some 1000 roubles worth (there are from 1000 to 2000 poods of blubber), forms the clear profit of the whaler. At present, there is a company, with a considerable capital, beeng started in St. Petersburg, which intends, next year, to start whaling along the Mourman coast.

We have no information as to the number of whales in the eastern part of the Arctic and in the Berring Straits.

Putting aside the products, got by the inhabitants of the Arctic coast, which, at any rate, is of some consequence, and only counting the products of regular whaling seal fishing, we remark the very extraordinary fact, that the wide-spreading Arctic Ocean, with its many gulfs, and the White Sea, yield a great deal less than the smaller Caspian does by nothing but its seals. As there are more animals (even seals) than in the Caspian, this can only be accounted for by the thorough way in which the business is carried on in the Caspian, where it is aided by natural conditions, by the comparative ease of killing seals, and by the presence of capital and enterprise. In the north, on the contrary, the danger and difficulty of the trade, and the absence of a population, counteract the possibility of its yielding as great a quantitis of useful products as it might well do, without destroying the natural abundance.

In consequence of this, one cannot help wishing that whaling etc. would increase in the north, and that more care would be taken in sealfishing on the Caspian, where seals may be completely exterminated in a considerably short time. We may remark that — *as many very valuable animals, for ex. the Greenland whale, Kamtchadal otter etc., are gradually dying out, and are in danger of the fate of their cousin, the sea-cow (Rhytina Stelleri), and as it is next to impossible for one state to prevent it, it is very desirable that a committee should be formed, for the working out of a set of rules for hunting, trapping etc., which would be binding on all countries.*

From the above it will be seen how difficult it is to determine the exact quantity of useful products yielded by whaling, fishing, trapping, etc. One can only make a guess that the sum total of the blubber yielded by the seal and other fisheries in the White Sea, Baltic, Lake Ladoga and the Caspian, reaches, on an average, about 150,000 poods, for the sum of about 500,000 roubles, the figures given by the traders always being below the real number. As regards the export of blubber and oil from Archangel, it is very small.

XI. The Invertebrates of the water fauna, which are objects of trade.

THE RIVER CRAWFISH.

There are six kinds of crawfish in Russia, the geographical distribution of which is highly interesting.

In European Russia there are four kinds;

1. The b r o a d - c l a w e d c r a w f i s h (*Astacus fluviatilis Rondelet*) is a western species, and in Russia it is found only in the basin of the Baltic, to where it emigrated from the west after the glacial period, and has now spread far to the north and to the east. In Finland it is found as far north as Christianstadt (62° 16′ n. lat). Then as far as Serdobol (62° 42′ n. lat), where it has been brought lately. Towards the east it is found in the rivers falling into the Ladoga Lake (Svir, Msta, Volkhoff) and in the south and central parts of the Baltic basin. Besides these places, it has been artificially cultivated near the source of the Dnieper. But in the east it has been gradreally driven out by

2. The l o n g - c l a w e d c r a w f i s h (*Astacus leptodactylus Esch.*), which is spread over the Caspian and Azoff seas, and in the fresher parts

4

of the Black Sea (i. e. those that do not contain more than 1,3⁰/₀ of salt) and also in all the rivers falling into these seas. From these river it has spread along the N. Dvina to Archangel, also into the Baltic system and is met with in the r. Volkhoff, Msta. Svir, etc. This crawfish has the remarkable power of adapting itself to its surrounding, and for this reason, many local varieties have been formed. In the Caspian it is not more than 12 centimetres long, but in the Volga, at Astrakhan, Saratoff, and particularly at Samara, it reaches the great length of 40 cent. I have heard that some time ago there was such a crawfish, 70 cent. long, for sale in St. Petersbnrg; it was preserved in spirits. Though the long claw-ed crawfish is fatter, it is considered less tasty than the broad-clawed crawfish, though the flesh of the latter is not as tender.

We shall speak further on of the acclimitization of the *Astacus lepto-dactylus* in Siberia.

3) Thick-clawed crawfish (*Astacus pachypus Rathne*) is most like the *Astacus fluviatilis,* from which is has most likely deve-loped, having found its way into the Chersono-Caspian basin at the time that the communication of this basin with the Arctic (along the Ural) was broken off, and when the Aral Sea was a salt bay of the Caspian (like Karaboogaż at the present day). Being quite a brackish-water form, this crawfish never goes very far up the rivers, keeping near the mouth, and in the Caspian Sea at the depth of 17 fathoms.

For this reason the *Astacus pachypus*, with some representatives of the genus *Gobius* and *Benthophylus*, characterize the Caspian basin of the postpliocenic period.

4) The Colchis-crawfish (*Astacus colchicus Kessl.*) which lives, as far as is known, in the basin of the r. Rion; it is a form between the *Astacus fluviatilis* and *Astacus pachypus*, and is about 13 centm. long.

In Asiatic Russia, crawfish are found only in the east, namely:

5) Daurian crawfish (*Astacus dauricus Pallas*), which is found in the upper part of the basin of the r. Amoor, in what is called Dauria; and 6) Schrenk's crawfish (*Astacus Schrenkii Kessl.*), which lives in the middle and lower parts of the r. Amoor. In the rest of Asiatic Russia, not excepting even the Aral Sea with its tributaries, there are no river crawfish, or at least, there were none in the last century, but they have found their way through the system of the Ural into the r. Tobol and its tributaries. They were also partly introduced by artificial means, at the beginning of this century, in the twenties.

In 1821, for the first time, 200 crawfish were brought by Mr. Jetisoff from the river Tchoosova· (a tributary of the Kama) and let out into the r. Iset (a tributary of the Tobol)· where they multiplied, so that on the third year they attracted the attention of the natives. Afterwards, from the Iset, crawfish were taken to other rivers, as, for example the Archimandrite Vladimir and Mr. Slovtsoff let out 20 cart-loads of Iset crawfish into the river. Toora. At the present day, the acclimatized crawfish in Siberia occupies an area from 55°—60° n. lat. and from 66°—78° east long.

Wherever the crawfish is met with, it is found in great numbers; this is partly owing to free life and the great quantity of food in the water, and partly owing to our peasants not eating crawfish, and therefore, not catching them. There have been cases, where in some districts the peasants willingly ate crawfish, as long as they considered them to be fish, but when they found out their mistake, they would not eat them any more, considering them unclean. For this reason, crawfish are not caught in most places, and in towns, where they are eaten by the upper classes, they are merely crawfish that have been caught along with the fish. In the villages, these chance crawfish are sometimes (though seldom) given to the pigs. In consequence crawfish are almost of no value, and at the fisheries anybody is welcome to take them gratis from the nets, when the fish is being taken out. Even in well populated towns, like Samara, Saratoff and Astrakhan, in the market one can buy a hundred crawfish for some ten or fifteen copecs.

On the other hand, in places where crawfish are much eaten, as, for instance, in Moscow, St. Petersburg, and the towns of the western governments, the prices are very high; for example, in St. Petersburg, crawfish from 14 to 18 centim. long are sold at 6 —8 cops. a piece. This has, of course, caused the importation of crawfish from more distant places.

They are brought to St. Petersburg by rail from the neighbouring governments, namely, from the Ostashkoff and Vishni-Volotchek districts of the government of Tver, and the Valdai and Novgorod districts of the government of Novgorod, and also from Ostroff (government of Pskoff), The crawfish are packed in large baskets in wet straw. It we are to believe M-r. Popoff, 70,000 rbles' worth of crawfish are brought to St. Petersburg (this is very likely, as in 1881 6283 roubles' worth were exported from Finland alone). This will make it about 2.500,000 crawfish. Besides this, at Aleshki, a place near the mouth of the Dnieper, the peasant women prepare an original sort of preserves out of the tails of the crawfish, by drying them in ovens and boiling them down. These conserves are very nice, and could bring a good deal of profit, if they were more universally known. At Aleshki, these crawfish-tails are sold at 7 or 8 rbles per pood. It is a pity that as yet no one has taken either to exporting crawfish, or to canning them, like lobsters. There is no doubt that it would be very profitable, as it requires very little capital, and crawfish are so very cheap here. According to M-r. Popoff, at Aleshki a woman can earn from fifty to sixty copecs a day by catching crawfish, selling them at the rate of a copec a hundred, so that the daily catch must be about 5,500.

MOLLUSCS.

Of all the molluscs found in the waters of Russia, only two species are caught for sale, and these not much; these are 1) the pearl oyster, and 2) the common oyster.

1. The Pearl oyster (*Unio margaritifer*) and its pearls.

The pearl oyster is spread all over European an Asiatic Russia, particularly in the rivers of the north, with their clear, transparent water, as, for ex., the r. Siuzma (Arctic Ocean) and Soosha (White Sea), Poventchanka and Nemen (Onega) are famous for pearls-oysters. But they are also found in the rivers of the other basins, the Baltic, Black Sea (the systems of the Dnieper and the Don), the Caspian (the Volga system), and sometimes yield very good pearls.

Pearls have been fished for by the Russians time out of mind, and in considerable quantities, to judge by their being so frequently used, in such great numbers, in the Russian costume, to ornament images and church utensils. To the present day, in many places (f. ex. in the governments of Tver, Novgorod, Olonetz, Vologda and Archangel) the merchant's wives and well-to-do peasant-women wear pearls on their head-dresses and earrings. In 1721, by the representation of the Mining College (or Ministry), an Ukaz was issued forbidding any private person, even the landowners themselves, to fish for pearls in the government of Novgorod, and in the Rzhoff and Toropetz districts of the government of Tver. This Ukaz also gave some instructions, as that 1) pearls-fishing to go on only from the middle of July till the midle of August; 2) to look for pearl-oysters in fresh, clear water, especially in places where the gudgeon, uklei and river trout are found (2 clauses, published in the Code of Laws v. XII, §§ 753 and 754). But the following year this prohibition was done away with, and replaced by a supervision of free pearl-fisheries. Only fully grown pearls were to be taken, and the larger pearls were to be sent to the College of Commerce (for a remuniration). Then, November 9-th 1766, some more rules were issued for the preservation of pearl oysters; it is interesting to read the following rule (Code of Laws, v. XII, § 757): «The fisheries must leave some places in the rivers, from which no oyster are to be taken, so that the latter should not be completely exterminated».

It is impossible to say anything about the exact state of pearl-fishing at the present day. It is only known that in some places (chiefly in the northern governments) the peasants fish for pearls, and sometimes come across valuable ones, which are often passed off by jewellers as real sea-pearls, and that there are firms (in Moscow) which deal exclusively in jewels made of Russian pearls. At any rate, considering, on one hand, the multitudes of pearl-oysters found in Russia, and on the other hand, the absence of well-conducted pearl fisheries, there must be a very great number of pearls in our numerous rivers. About 2000 roubles'worth of unset pearls are yearly exported from Russia. — The *Unio plicata* (of such consequence in China) is probably to be found in the Amoor.

Besides the *Unio margaritifer*, pearls are got from the *Mytilus latus Chemn.* Some years ago M-r Stribnitzky brought (from Theodosia, on the Black-Sea) a *Mytilus latus*, in the shell of which some 150 small pears had been discovered.

2) The oyster (*Ostrea adriatica Lamk.*) is found in considerable numbers along the coast of the Crimea. Though small, it is very tasty, and is the object of a considerable trade. We may remark, by the bye, that in Kertch, oysters brought from Theodosia cost 40—50 copecs a 100, and in St. Petersburg 10 of them cost 50 copecs.

As to the laws connected with oyster-fishing, we only know of one (§ 773, v. XII of the Code of Laws). «For taking small oysters on shore from the Theodosian fisheries, a fine of 1 rouble 50 copecs is to be imposed for each offense».

Besides oysters, several other molluscs are eaten, but only by the inhabitants of the coast of the Black Sea. These eat (either raw or boiled) the *Mytilus latus*, the sea-comb (*Pecten sulcatus*) and various kinds of *Venus* and *Cardium*.

LEECHES

(Hirudo officinalis, var. bicatenata. H. colchica).

The medicinal leech is spread all over Russia, being met with in the governments of St. Petersburg, Novgorod and Olonetz, and all the governments to the south of these. The number of places infested by leeches of course increases towards the south; but the real land of leeches is Trans-Caucasia, viz. the districts of the Black Sea, Poti and Lenkoran. The deeply-shaded rivers and forest - bogs of the Lenkoran district regularly teem with leeches, so that it is impossible to bathe there. In every net full of ooze one draws up, some 20 or 30 leeches are sure to be found. In the forties, fifties and sixties of this century, when the use of leeches in medicine had reached its height, when the leeches in the Parisian hospitals alone sucked out 90.000 kilogrammes of human blood, and when 7,000.000 leeches were not enough for the London hospitals, — the demand for leeches, and therefore the sale of them, was very considerable in Russia. As the central and northern governments had not leeches enough of their own, the latter were brought (to Moscow as a centre) from Bessarabia, Astrakhan and Trans-Caucasia. It is true, a certain quantity of leeches was imported from Hungary, but then the Lenkoran leeches were exported.

How great the sale of leeches was, can be judged (not having any statistics) by the fact that many Trans-Caucasians (chiefly the sectarian exiles) enriched themselves by exporting leeches.

However great our natural supply of leeches might have been, it was apparently too small to satisfy the demand for them. So, on one hand, special orders were issued by Government (§ 562 v. XII C. of L.) ‹rules for catching leeches in ponds and lakes› (issued Sept. 21, 1848), which, by the bye, 1) forbade leeches to be caught during May, June and July, and 2) to take leeches either of too small a size (not less that $2^5/_8$ inch.) or, large, old leeches, which were not fit for medicinal purposes, but only for propogation of the species, and 3) recommended the breeding of leeches.

And on the other hand, leech-breeding establishments were started (apart from the above recommendation). Artificial leech-ponds (or *parks*) were built on Sauvet's system, for instance, in Moscow (M-r Parman), in St. Petersburg (M-r Gavriloff), in Piatigorsk, in Nijni-Novgorod, on the Ural (by Malysheff, the pisciculturist of 1855).

Soon, however, all these measures became unnecessary. Doctors repudiated leeches. Yesterday, leeches were benefactors, to-day—they were dangerous and harmful, and were therefore left in peace in their native bogs.

SPONGES.

Of all the spongilla, only the fresh-water sponge, known under the name of the *badyaga*, is an article of commerce, and even that not much. There are four kinds of *badyaga* — *Spongilla lacustris, Spongilla sibirica, Tranchispongilla Mülleri* and *Ephydatia fluviatilis*. In various places in Russia (chiefly in the central and southern governments, where during the long summer the *badyaga* can attain a considerable size), the peasants gather and dry the *badyaga* and sell them to the druggists. This trade is pretty extensive in the governments of Kiev, Poltava, Kharkoff, and the neighbouring ones.

Published by the order of the Ministry of State Property.

CPSIA information can be obtained
at www.ICGtesting.com
Printed in the USA
BVHW041307281118
534217BV00007B/25/P

9 781528 463492